STEEL CITY

Hamilton and Region

From its establishment nearly 200 years ago as a village at the centre of an agricultural district, Hamilton has grown into one of Canada's biggest industrial centres, at the heart of a highly developed regional municipality. The story of its changing landscapes, both physical and human, is presented in the nineteen essays that make up this volume, all by geographers associated with Hamilton's McMaster University.

Change is the essence of the story. Each contributor focuses on one aspect of the past, present, or future landscapes of Hamilton, and places it within the context of change in the region. The first series of essays explores physical landscapes – geology and relief, climate, soils, vegetation, and hydrology – and shows how human activity has moulded them.

The second group charts the evolution of human landscapes in the region, paying special attention to contemporary Hamilton with its rich and diverse combination of people and cultures, and also to the political intrigue that surrounded the introduction of regional government to the area.

Finally a third series focuses on the functioning of the Hamilton region. Within a highly complex system, the city and region balance a broad range of often contradictory trends and activities. The contributors examine the difficulties facing agriculture in a rapidly urbanizing region; the importance of Hamilton in caring for welfare-dependent populations; the future of steel in Steel City; the challenges posed by energy requirements in the region; and the hard choices facing policy-makers. The last two essays discuss the role played by McMaster University in the life of the region, and the landscape of Hamilton today: a remarkable complex of historical interest, great natural beauty, and modern city life.

M.J. DEAR is a member of the Department of Geography at University of Southern California. J.J. DRAKE and L.G. REEDS are members of the Department of Geography at McMaster University.

STEEL CITY

Hamilton and Region

Edited by

M.J. Dear, J.J. Drake, and L.G. Reeds

UNIVERSITY OF TORONTO PRESS
Toronto Buffalo London

© University of Toronto Press 1987
Toronto Buffalo London
Printed in Canada

ISBN 0-8020-2563-3 (cloth)
ISBN 0-8020-6582-1 (paper)

Canadian Cataloguing in Publication Data
Main entry under title:

Steel city

Includes index.
ISBN 0-8020-2563-3

1. Physical geography – Ontario – Hamilton Region.
2. Anthropo-geography – Ontario – Hamilton Region.
3. Hamilton Region (Ont.) – Description and travel.
I. Dear, M.J. (Michael J.). II. Drake, J.J. III. Reeds, Lloyd George, 1917-

GB132.05S78 1987   917.13′52   C87-093749-9

The cover photograph, 'Hamilton Centre,'
is reproduced with the kind permission of the photographers
Neil S. MacPhail / John Bruce (Hamilton).

THIS BOOK IS DEDICATED

TO THE PEOPLE OF HAMILTON

AND ITS REGION

The essays collected here honour the more than fifty years of association between McMaster University and the City of Hamilton and the celebration of McMaster's Centennial year (1887-1987). The volume is also intended to mark the fortieth anniversary of McMaster University's Department of Geography.

Contents

II. THE BUILT ENVIRONMENT

III. HOW HAMILTON WORKS

Figures

Tables

Foreword

Seated in the middle of the McMaster University campus are two fine neo-Gothic stone buildings that have come to symbolize McMaster. One of these, Hamilton Hall, is named after the city which, back in the 1920s, invited the university to come to it from Toronto. Even though the Great Depression was just beginning, the people of Hamilton raised half a million dollars to finance the handsome building that still stands as a monument to the generosity and intelligent self-interest of this community.

One of the university's outstanding achievements in the years since 1930 has been the creation and development of our Department of Geography. Now one of the strongest geography departments in North America, it has marked its fortieth anniversary in 1986 by preparing this handsome volume in time for McMaster's Centennial in 1987. The publication is a very fitting tribute to the city and region that have been so excellent a home for McMaster during its Hamilton years.

Alvin A. Lee
President and Vice-Chancellor
McMaster University

Avant-propos

Au centre du campus de l'université McMaster, se trouvent deux édifices de pierre, magnifique spécimens de l'architecture néo-gothique, qui en sont devenus le symbole. L'un d'eux, Hamilton Hall, porte le nom de la ville qui a invité l'université à venir de Toronto pour s'y établir dans les années 20. Les habitants de Hamilton, malgré la Dépression incipiente, réussirent à rassembler un demi-million de dollars pour élever ce superbe édifice qui sert toujours de monument à leur générosité et à leur prévoyance éclairée.

La création et l'expansion de notre département de géographie (maintenant l'un des meilleurs de l'Amérique du Nord) a été l'un des accomplissements les plus mémorables de l'université depuis les années 30. Le département a commémoré son quarantième anniversaire en 1986 par la préparation d'un élégant volume qui paraîtra lors du centenaire de McMaster, en 1987. Cette publication rend un juste hommage à la cité et à la région qui ont offert à McMaster un lieu de résidence incomparable depuis son arrivée à Hamilton.

Alvin A. Lee
Président et Vice-chancelier
L'université McMaster

Acknowledgments

The publication of this book has been made possible through grants from McMaster University's Centennial Committee, the Office of the Vice-President (Academic), and the Department of Geography.

We are grateful for the support and encouragement of the staff of the University of Toronto Press (especially Mr Virgil Duff).

Richard Hamilton drew all the illustrations for this volume.

The manuscript was prepared by staff of the McMaster University Faculty of Science Word Processing Centre. The cover photo is reproduced by permission of Mr Neil S. MacPhail.

M.J.D., J.J.D., L.G.R.

INTRODUCTION

CHAPTER 1

Hamilton: region in transition

M.J. DEAR, J.J. DRAKE,
AND L.G. REEDS

This book is a celebration. Almost two hundred years ago, a number of survey lines were drawn, setting in motion the series of events that would lead to the establishment of the city of Hamilton. One hundred years later, the city was a thriving industrial metropolis. English visitors in 1884 wrote the following tribute: 'Of all the places we had visited during our trip to the American continent, the prettiest, cleanest, healthiest, and best conducted was the City of Hamilton, Canada.' It was also one hundred years ago that McMaster University was founded (in Toronto). Fifty years after its founding, the university moved to Hamilton, and the destinies of both city and university have been inextricably linked since then.

During the past fifty or more years, there have been a number of major social upheavals, including an economic depression, a world war, and a technological revolution. Both city and university have experienced the ebb and flow of these changes – the good times and the bad times. Yet both Hamilton and McMaster continue to flourish as vital components in the national, provincial, and local scene. How have we endured? What kinds of changes have we experienced in this region? And how have 'town' and 'gown' adapted to these changes?

In this book, we look at the changing fortunes of Hamilton – its past, present, and future. We believe the outcome of this assessment provides a cause for celebration, not because all our problems have been solved, but because many past difficulties have been successfully overcome. Together, the people of Hamilton and McMaster have created one of the best places in Canada to live, work, play, and study. We hope that when you have read this book, you will know more about Hamilton and its region and will join its peoples in this celebration.

Birth, growth, and decline are inevitable features of the human and physical landscapes. For example, in the human landscape, changing economic and political fortunes can cause a relatively rapid decline in once-powerful nations and cities; and, in the physical landscape, prominent landscapes are ultimately worn down to featureless plains through the persistent forces of erosion by wind and water. In Hamilton, we have witnessed many changes during the past decade in patterns of urban growth, employment, and land use (especially in agriculture). But our research has also uncovered many constants that testify to our ability to persist, to withstand the onslaught of change.

Most human life and natural phenomena seem to be guided by distinct rhythms, or cycles, in their evolution. In human activities, researchers have observed that much economic change occurs in the form of regular cycles. Some of these cycles are very long, in temporal terms; the depression of the 1930s and the recession of the 1980s are usually regarded as parts of a long-recognized fifty-year economic cycle of boom and bust. Other 'business cycles' can be much shorter. It is also important to know that cycles occur at many different scales of human activity. Figure 1.1 shows the long-term cyclical pattern in the Canadian wholesale price index (a); the more rapid fluctuations in the rate of unemployment in Ontario (b); and the regular variations in the vacancy rates in hospital beds at the Hamilton Psychiatric Hospital (c). Another cyclic pattern describes Hamilton's rate of population growth (figure 1.2).

The natural environment is similarly characterized by a variety of rhythms. The hundred-year cycle of precipitation in Hamilton is shown in figure 1.2, and the city's seasonal temperature variation is shown in figure 1.3. While the rhythms and cycles which underlie the natural environment may take much longer to become visible to human eyes, they, none the less, make the physical landscape an active and vital one. Natural processes have had, and continue to have, a significant impact on the development of Hamilton and its region. Equally important, however, is human-induced change in the physical landscape. Natural rhythms in landscape evolution are often interrupted by human practices such as farming and urban development. This interruption can, in many cases, produce disastrous consequences. In the Hamilton area, the most urgent examples of human-induced environmental degradation are the pollution of the Great Lakes and the impacts of acid rain.

In this book, we look at growth and change in both the physical and the human landscapes of Hamilton. We examine the past, present, and future to provide a comprehensive account of the modern city and its region. As you read with us, we shall come to see that Hamilton has gone

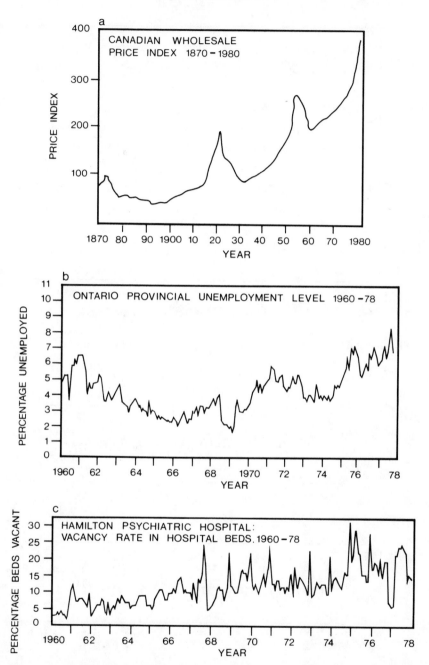

Figure 1.1 Trends and cycles: prices, unemployment, and hospital vacancies

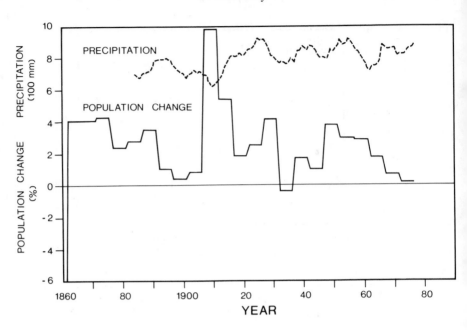

Figure 1.2 Human and physical cycles: population change and precipitation

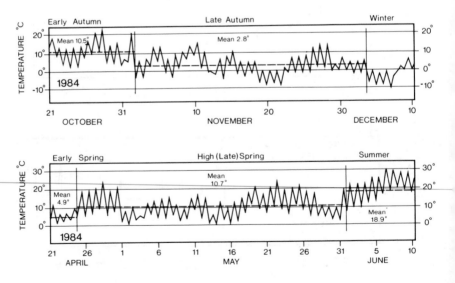

Figure 1.3 The city's seasonal temperatures

through many changes and upheavals during its history. The city has survived these transitions, making constant adjustments to fit the changing times. Looking ahead, we are confident that history will repeat itself; that Hamilton and its peoples will weather the current round of social and economic changes as they prepare for the challenges of the twenty-first century.

OVERVIEW OF THE BOOK

The essays in this book are grouped in three parts: an examination of the natural environment and the built environment, and an account of the city and region at work.

The Natural Environment

The five essays in part I are concerned with the evolution of the physical landscape of Hamilton and the significance of that landscape for human activity in the region. Our contributors focus on the landforms (chapter 2), climate (chapter 3), soils (chapter 4), and vegetation of the region (chapter 5). They provide glimpses of past and present landscapes, as well as discussions of future threats to our environmental resources. The impact of human activity on the natural environment is the central concern of chapter 6, which shows how human interference can significantly disrupt the hydrology of our regional landscapes.

The Built Environment

The four essays in part II examine how and why the City of Hamilton and its region developed at the western end of Lake Ontario. The beginnings of Hamilton are reconstructed in (chapter 7) from the evidence of five nineteenth-century maps. Then, in chapters 8 and 9, the twentieth-century growth of Hamilton is described, with particular emphasis on the social and ethnic diversity of Hamilton's population. Finally, no account of Hamilton would be complete without a discussion of the well-publicized enforced amalgamation of Hamilton and its neighbouring municipalities which led to the creation of the Regional Municipality of Hamilton-Wentworth (chapter 10).

How Hamilton Works

The character and personality of a region depend on how people and resources come together to work. By 'work,' we do not simply mean what goes on behind the factory gates; rather, we are referring to the complex texture of everyday life in Hamilton. In this sense 'work' involves teaching the young, caring for the old, nurturing culture, and creating a community. In the seven essays in part III, our contributors consider several aspects of what makes Hamilton work. The role of Hamilton as the 'steel city' of Canada is examined in chapter 13, and our contributors discuss some of the tough choices Hamilton faces in planning and development decisions (chapter 15) and in land-use policy for agriculture (chapter 11). Part III also addresses issues of which we may not be so aware: the role the city plays in caring for its welfare-dependent populations (chapter 12); the difficult energy choices facing Hamilton (chapter 14); and the vital role played by McMaster University in the city's economic, social, and cultural life (chapter 16).

As we look at Hamilton today (chapter 17), the diversity and resilience of the city and its region, we witness an impressive human achievement. Rapid changes are well underway in the region; a new city is being built on the foundations of the old. Yet these changes are enacted against a backdrop of rich history, formed from a strong social fabric. Hamilton has always had a proud record of ethnic diversity and interaction; Hamilton will remain a region of intense physical beauty. In this book, we will examine the changes occurring in our region, and explore what the future may hold. In doing so there will emerge a fitting tribute to the enduring traditions in the physical and human landscapes.

PART ONE

THE
NATURAL
ENVIRONMENT

Right, top: Hamilton Mountain
Right, bottom: Tews Falls
(photos by R. Bignell)

Cootes Paradise
(photo by R. Bignell)

Westover drumlins
(photo by L.G. Reeds)

An esker used for gravel
(photo by R. Bignell)

CHAPTER 2

Physical landscape of the Hamilton region

S.B. McCANN

Hamilton is situated in the eastern part of the Great Lakes basin of North America, at the head of the lowest of the Great Lakes, Lake Ontario (figure 2.1). The lake surface is 74 m (243 ft) above sea level (asl) but the highest point in the Hamilton-Wentworth region, only 24 km (15 mi) from the lakeshore, is more than 300 m (975 ft) asl. Thus, although the region is located within lowland terrain, essentially an eastward extension of the St Lawrence lowlands, it is characterized by considerable local relief and a variety of landforms.

The physical landscape of the Hamilton area is dominated by the steep, often vertical, outcrop of Silurian rocks in the Niagara escarpment, which is known locally as Hamilton 'Mountain.' The common and affectionate use of the term 'the Mountain' to describe either the steep escarpment face or that part of the city above the escarpment, and its official use as the name of a federal parliamentary constituency (Hamilton Mountain), reflect the importance of this feature of the landscape in the minds of the people of the region. Away from the escarpment, bedrock is mantled, for the most part, by a variety of unconsolidated glacial deposits and other surficial materials. The Silurian rocks were formed more than 400 million years ago in warm tropical seas: the glacial drift was deposited in the late Pleistocene epoch, about 12,000 to 13,000 years ago, during the final retreat of the Ontario lobe of the late Wisconsin Laurentide Ice Sheet. During the last 10,000-11,000 years, known as the Holocene epoch, streams and slope processes have eroded the soft unconsolidated drift, producing steep ravines and dissected topography, and mass movement processes (rockfall, landslips) have modified the face of the escarpment. During this period also, soil development took place and the vegetation, which initially colonized the barren landscape left by the glaciers, developed into the deciduous and mixed forest found by the European explorers and removed

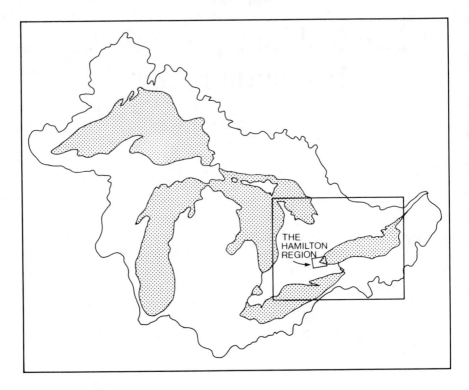

Figure 2.1 The Great Lakes Basin, showing the location of the Hamilton region. The larger box shows the area contained by figure 2.2

by the early settlers. The lake in the Ontario basin, which acts as a local base level for streams, was first higher and larger, and then lower and smaller, than the present Lake Ontario, and these changes are also manifest in the landscape.

The city of Hamilton has a fine natural setting and there is considerable scenic appeal in the surrounding countryside. Many of the best locations are preserved and maintained as conservation areas or within the lands of the Royal Botanical Gardens. This chapter seeks to explain the evolution of the interesting and attactive physical landscape of the region and considers some of the the geomorphic processes which have been important in this evolution. The chapter begins with a discussion of the bedrock geology and the nature of the escarpment. Then the late Quaternary history of the region and the formation of the suite of glacial landforms are considered, followed by a discussion of the various lakes which have existed in the Ontario basin and a description of the shoreline features of the modern lake. Stream processes and products, ravines and flood-plains, are considered last, together with the hydrological regime, or flow pattern, of the modern streams.

It is useful to consider the physical landscape of the Hamilton region in the context of the physical landscape of Southern Ontario, for certain aspects of the local scenery are common throughout the broader region. Figure 2.2 illustrates this point. Figure 2.2(a) indicates that the rocks which underlie the Hamilton region and which outcrop in the escarpment occur as a broad belt across Southern Ontario from Niagara through to To-bermory. The Queenston shale sites at the base of the escarpment, the Clinton and Cataract groups of rocks occur in the escarpment face, and the dolomites of the Guelph and Lockport-Amabel formations provide the cap rock. The Niagara escarpment is the principal landform not only of the Hamilton region but also of Southern Ontario. Figure 2.2(b) shows the regional pattern of glaciation some 13,000 years ago when glacial ice, as the Ontario lobe, filled the Ontario basin and extended southwestwards to prominent terminal moraines, the Paris and Galt moraines, which de-fine the western limit of our study area. Farther north, at the same time, ice of the Georgian Bay–Simcoe lobe extended southwards to a position above the escarpment where similar terminal moraines, the Singhampton and Gibraltar moraines, were deposited. The two ice lobes were separated by the Oak Ridges moraine. The pattern depicted in figure 2.2(b) repre-sents one stage in the retreat and wastage of the great Laurentide Ice Sheet which reached its maximum extent, covering most of eastern and central Canada, about 22,000 years ago and had wasted to residual ice domes either side of Hudson Bay by about 7,000 years ago. As the ice margin retreated northwards and eastwards through the Great Lakes region, huge volumes of meltwater were dammed between the retreating ice and the higher ground around the lake basins, creating a complex series of short-lived glacial lakes. The early lakes drained southwestwards into the Mississippi, but by the time the Ontario basin was clear of ice, and Lake Iroquois was dammed by ice across the St Lawrence exit, the drainage was to the southeast via the Hudson Valley to the Atlantic. As figure 2.2(b) shows, similar suites of glacial and related deposits and landforms to those of the Hamilton region occur throughout Southern Ontario.

BEDROCK GEOLOGY AND THE ESCARPMENT

Figure 2.3 is a diagrammatic cross-section of the upper part of the Niagara escarpment in the vicinity of the Jolly Cut in Hamilton. It shows the relative

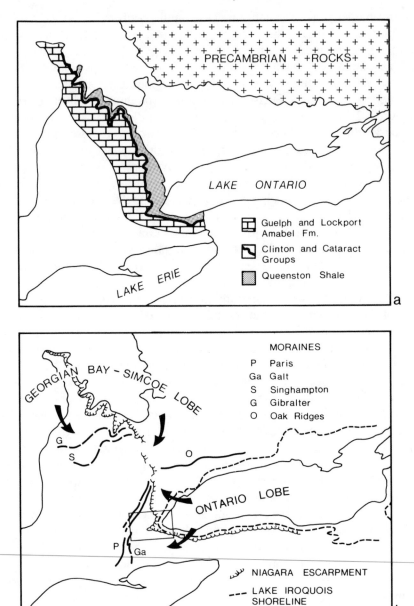

Figure 2.2
(a) The regional extent of the rocks which outcrop in the Hamilton region
(b) The regional pattern of glaciation about 13,000 years ago when the
Paris-Galt moraines were formed

thickness and lithology of the strata which are exposed and indicates the form of the escarpment face. A similar sequence of rocks constitutes the escarpment throughout Southern Ontario, but there are lateral changes in lithology, and different names may be used for similar strata in different localities (Bolton, 1957). The most important of the rocks in the escarpment is the Lockport dolomite, which in Hamilton, as in most of the area, forms the prominent vertical face below the escarpment crest. At the Jolly Cut, two members of the Lockport formation occur: the Goat Island (Ancaster) member, a thinly bedded, well-jointed dolomite, with chert nodules, and the Gasport member, a more massive dolomite. The rocks below the Lockport formation, down to the Queenston shale, which occupy the face of the escarpment, constitute the Clinton and Cataract groups. The narrow outcrop of these rocks, as shown in figure 2.2(a), defines the outline of the escarpment across Southern Ontario. They consist of less resistant, thin sandstones, shales, dolomites, and limestones, though individual beds and some whole formations can be very resistant to erosion at particular localities. At the Jolly Cut the Irondequoit formation is a 2 m (6.5 ft) thick, single bed of resistant dolomite, which stands out between the recessed Rochester shales and the weaker Reynales formation. Near the bottom of the section the Whirlpool sandstone is another resistant formation. The Queenston shale at the base is a very distinctive red shale with streaks of green, which includes thin siltstone, fine sandstone, and calcareous interbeds. The Queenston shale, which occupies all the lower ground below the escarpment, is usually mantled with surficial deposits but it is exposed along some of the major creeks. Good exposures of the rocks of the escarpment, particularly from the Reynales formation to the Ancaster member, may be seen at all the major road cuts on the highways leading southwest and northwest from the city: Highway 20, Highway 403, Sydenham Cut (Dundas), Clappison's Cut (Highway 6). Fine, natural cliff sections occur at Dundas Peak and Mount Nemo, and at various waterfalls over the escarpment. The escarpment crest in Hamilton is 180-185 m (590-607 ft) asl or some 110 m (360 ft) above the level of Lake Ontario. The crest rises northwards to 290 m (950 ft) at Mount Nemo.

The arrangement of the moraines and other glacial deposits in Southern Ontario leaves no doubt that the escarpment was a major feature of the landscape before the last stages of the Pleistocene glaciation. However, the form and significance of the escarpment prior to the onset of glaciation are matters of some debate, which also involves views about the nature of the pre-glacial river system. A comprehensive treatment of this problem, which focuses on the major re-entrants, or valleys, which break the con-

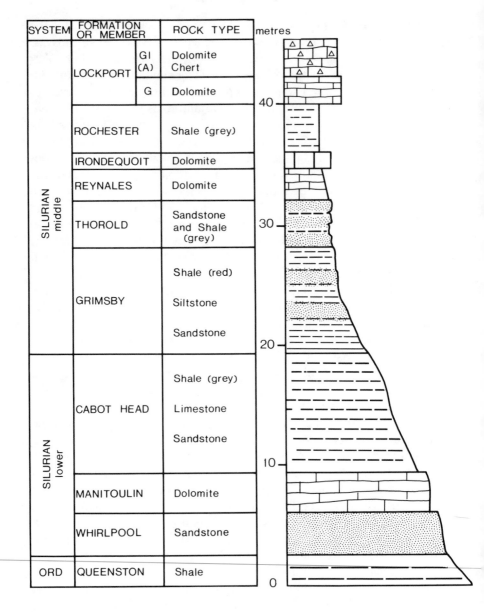

SYSTEM	FORMATION OR MEMBER		ROCK TYPE	metres
SILURIAN middle	LOCKPORT	Gl (A)	Dolomite Chert	
		G	Dolomite	40
	ROCHESTER		Shale (grey)	
	IRONDEQUOIT		Dolomite	
	REYNALES		Dolomite	
	THOROLD		Sandstone and Shale (grey)	30
	GRIMSBY		Shale (red) Siltstone Sandstone	20
SILURIAN lower	CABOT HEAD		Shale (grey) Limestone Sandstone	10
	MANITOULIN		Dolomite	
	WHIRLPOOL		Sandstone	
ORD	QUEENSTON		Shale	0

Figure 2.3 Schematic geologic section of the Niagara escarpment at the Jolly Cut in Hamilton

tinuity of the escarpment, is given by Straw (1968a). His views are of particular interest where he provides an explanation of the origin of the Dundas Valley, which is one of the major re-entrants (12 km [7.5 mi] long and 4 km [2.5 mi] wide) in the escarpment. Straw noted that there are twenty-six similar re-entrant valleys along the escarpment between Niagara and Tobermory, and suggested that they were largely, if not wholly, produced by ice erosion during the Wisconsin glaciation. The Dundas Valley is interpreted as a major glacial trough carved by ice moving south-westwards from the Ontario basin. The Mount Albion Valley, at the eastern end of the city, is considered to be a smaller re-entrant formed in the same way. The Dundas Valley is remarkably straight and deep, but, because of the infilling of glacial drift deposited during deglaciation, the full extent and depth of the bedrock valley is not seen. Bedrock is well below modern lake level as far west as Copetown where the drift is 180 m (590 ft) thick.

Another feature of the escarpment in the Hamilton area is the occurrence of narrow gorges, headed by waterfalls, which extend back from the escarpment edge. Several of these are located on figure 2.4(a), as late and post-glacial gorges. The most striking is Spencer Gorge, above Dundas, which has two headward arms both with waterfalls, Websters Falls and Tews Falls. These gorges were initiated during deglaciation when large amounts of water were available from melting ice, and the present streams are underfit, though headward retreat by undercutting and collapse of the cap rock continues today. In a sense these waterfalls and gorges are miniature editions of Niagara Falls and the Niagara gorge.

The escarpment is not a static landform. The cliff face is subject to continual weathering and erosion, particularly in late winter when the effects of freezing and thawing are intense. The rockfall debris in spring is particularly apparent on the upslope verges of the mountain access roads: Sherman Cut, Jolly Cut, and the Queen Street hill. However, as with the process of stream erosion in the gorges, mass movement processes (rockfall, landslip, soil creep) on the escarpment were probably more active during and immediately after deglaciation. At the time, the bare rock and debris slopes left by glaciation would have been rapidly adjusting to more stable profiles.

LATE QUATERNARY HISTORY AND GLACIAL LANDFORMS

The Quaternary period, the last two million years of earth history, is dominated by numerous continental-scale glaciations. The landscape of

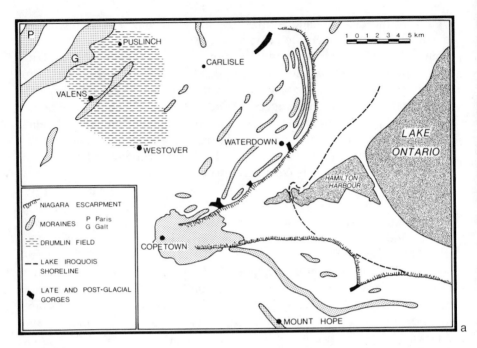

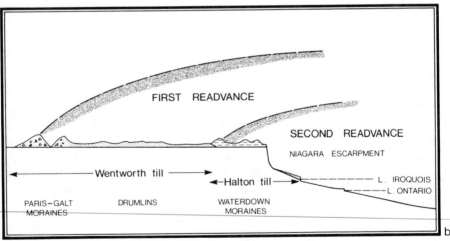

Figure 2.4
(a) Main geomorphic features of the Hamilton region
(b) Schematic cross-section from east to west across the region shown in (a) to
illustrate the extent of the glacial readvances out of the Ontario Basin which were
responsible for the glacial deposits of the region

the Hamilton region, and of Canada as a whole, cannot be understood without constant reference to the effects of glaciation. In particular, the effects of the last glaciation, the Wisconsin glaciation, are important, as each succeeding glaciation altered or destroyed the effects of the preceding ones. Glacial erosion throughout the Wisconsin scoured out the Ontario basin and carved the Dundas Valley re-entrant, and glacial deposition during the last retreat of the ice across the region produced a suite of depositional landforms.

The overall pattern of retreat of the late Wisconsin Laurentide Ice Sheet was broken by numerous readvances of the ice front, marked by the deposition of prominent terminal moraines. Two such readvances of the ice out of the Ontario basin were responsible for the glacial deposits and depositional landforms of the Hamilton region which have been mapped and described by Karrow (1963). The limit of the first readvance, which occurred about 13,000 years ago, is marked by two prominent moraine ridges, the Paris and Galt moraines, and a smaller ridge, the Moffat moraine (figure 2.4[a, b]). The Galt moraine is the most continuous feature. It is a prominent, hummocky ridge, 2-4 km (1.2-2.5 mi) wide, which is separated from the Paris moraine by a 3 km (2 mi) wide zone of outwash. The till deposited during the first readvance is a discontinuous, sandy, sometimes very stony, till known as the Wentworth till. Much of this till away from the moraines is in the form of drumlins – elongate, rounded, oval-shaped hills, which are streamlined landforms produced as the ice sheet advanced and moulded the till. The group of drumlins north of Westover is 1-2 km (0.6-1.2 mi) long, 300-600 m (1,000-2,000 ft) wide, and 23-30 m (75-100 ft) high (figure 2.5). These drumlins are aligned roughly east-west and are generally steeper at their eastern ends. Following the formation of the Moffat moraine the glacial advance ceased to be a dynamic event, and large masses of rapidly melting, stagnant ice were left above the escarpment as the ice front retreated into the Ontario basin. Straw (1968b) suggested that the meltwaters drained southeastwards, towards the escarpment, through and below the stagnant ice, and were responsible for the deposition of a series of eskers, the most prominent of which extended from Puslinch to Carlisle. Eskers are narrow, sinuous ridges of sand and gravel; because these materials are very suitable for construction purposes, most of the eskers of the region have been excavated and worked out.

The last late Wisconsin readvance of ice out of the Ontario basin did not extend far beyond the top of the escarpment. It deposited a series of low, sub-parallel moraines, the Waterdown moraines, north of the Dun-

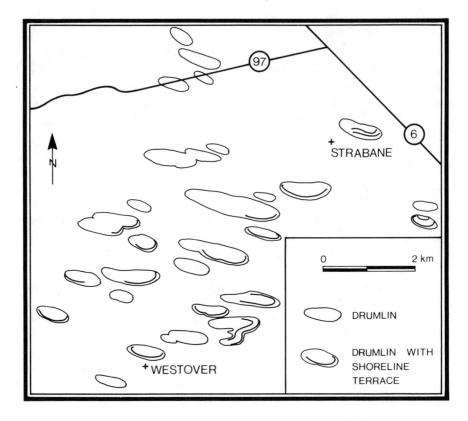

Figure 2.5 Drumlin field between Westover and Strabane

das re-entrant, and somewhat broader moraines to the south, towards Mount Hope. This readvance also formed the hummocky morainic and outwash terrain towards the head of the Dundas Valley, which buries the bedrock of the escarpment and extends as far west as Copetown. The till of this readvance is the red, silty Halton till, best examined east of the escarpment north of Burlington. It is grey and more clayey within the Dundas Valley, probably as a result of the incorporation of clays deposited in a temporary lake dammed in the valley by ice in the phase between the two ice readvances. The Halton till plain, north of Burlington, is gently undulating terrain developed on a till sheet up to 12 m (39 ft) thick, resting on Queenston shale. The till plain is characterized by flutes or grooves, which, like drumlins, are an indication of moulding by ice.

Evidence of erosion of bedrock by ice, rather than moulding of unconsolidated materials, is provided by the overall, trough-like form of the Dundas Valley, as suggested by Shaw (1968a), and by two other features at smaller scale. The first is the local abundance of large dolomite blocks, north and east of Carlisle, which must have been quarried by the ice from the escarpment edge and low bedrock ridges during the Wentworth till advance. The second is the glacially polished and striated character of the dolomite surface when this is newly stripped of overlying surficial deposits, particularly when the bedrock has been protected from weathering by clayey till. Good examples of glacially polished bedrock were exposed in excavations during the construction of Christies dam and reservoir on Spencer Creek.

Before considering the lake which developed in the Ontario basin as the ice front retreated northwestwards for the last time, reference should be made to the large lakes which developed in the Erie basin during and after the ice readvance which deposited the Paris-Galt moraines. The evidence that these lakes extended into the area encompassed by figure 2.4(a) is in the form of shoreline terraces which were developed along the southern sides of the drumlins near Westover and Strabane (figure 2.5). These shoreline features consist of wave-cut benches backed by low bluffs and wave-built, gravelly beaches. It is clear that at one stage the drumlins stood out as islands within a large water body which extended away to the southeast, and the logical correlation is with glacial lakes Whittlesey and Warren which were contained in the Erie basin by an ice front not very far west of the line of the Niagara escarpment.

GLACIAL LAKE IROQUOIS AND MODERN LAKE ONTARIO

The long-term history of water levels in the Ontario basin is shown in figure 2.6(a): figures 2.6(b) and 2.6(c) show the seasonal and cyclic fluctuations of water levels in modern Lake Ontario. Glacial Lake Iroquois developed about 12,500 years ago (Sly and Prior, 1985) and lasted for a few hundred years with lake level being controlled by the Rome outlet to the Hudson drainage. The prominent shoreline of this large lake has been described by Coleman (1937). In the Hamilton region the most striking feature of the shoreline is the large, gravel, barrier beach which separates Hamilton Harbour from Cootes Paradise. This narrow strip of land, which carries two highways and the railway out of the city to the north, was developed by the longshore transport of gravel from eroding bluffs to the north and south when Lake Iroquois stood 35 m (115 ft) above present lake level. The development of this barrier beach enclosed a lagoon to

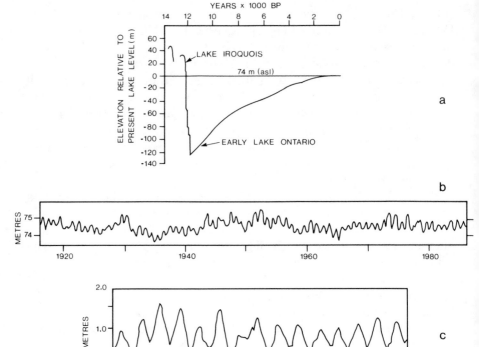

Figure 2.6
(a) Long-term history of lake levels in the Ontario Basin
(b) Fluctuations in the surface level of Lake Ontario from 1916 to 1985
(c) Fluctuations in the surface level of Lake Ontario from 1971 to 1985

the west in which silty sands accumulated. These materials now constitute the level ground between Westdale and Dundas. Elsewhere the shoreline is represented by a low erosional bluff parallel to and about 3 km (2 mi) north of the modern lakeshore in Burlington. In Hamilton a similar feature was mapped towards the foot of the escarpment but it is now obscured.

The shoreline plane of Lake Iroquois rises from about 35 m (115 ft) above modern lake level in the west to 120 m (394 ft) in the east, because of differential isostatic recovery. The eastern end of the Ontario basin was closer to the centre of the Laurentide Ice Sheet and subject to a greater ice load than the western end. Consequently the amount of crustal depression as a result of the ice load, and crustal recovery or uplift as the load was removed, was greater at the eastern than at the western end of the basin. When the ice cleared the St Lawrence outlet, bringing the Lake Iroquois phase to an end, the outlet was depressed and much lower than at present. Thus, lake levels fell drastically (figure 2.6[a]) and early Lake Ontario was a very small lake. The rise in water level over the past 12,000 years, to the modern lake level, reflects the continued isostatic uplift of the St Lawrence outlet. At the present time the eastern-outlet end of Lake Ontario is still rising with respect to the western-inlet end at a rate of 17 cm (7 in) per century. Consequently the shore at the western end of the lake is experiencing a gradual rise in water level.

Variations in precipitation are the main cause of the long-term variations in modern lake levels illustrated in figure 2.6(b), but there is no regular, predictable cycle between highs and lows. Lake levels were low in the 1960s, but high in the early 1970s, with a peak level in 1973, causing problems of increased shore erosion and flooding. There is a distinct annual cycle in the level of the lake with high-water level occurring in early summer because of inputs of spring, snowmelt runoff. The normal range from winter lows to summer highs is about 0.5 m (1.6 ft), but it can be up to 1 m (3 ft) in some years. Since the completion of the St Lawrence Seaway and Power Project in 1958, the outflow from Lake Ontario has been regulated. Outflow from the lake can be increased or reduced to alleviate some of the problem associated with high or low lake levels, but the natural controls still dominate the lake regime.

The shoreline of the modern lake in the Hamilton area consists of two elements. There are low bluffs in till and bedrock in Burlington and Stoney Creek which are linked by the sandy, barrier beach, known as the 'beach strip,' which separates the open lake from Hamilton Harbour. The contemporary equivalent of the Iroquois barrier beach described previously, the 'beach strip' has been built out from both the northern and southern shores of the lake by longshore transport of sediment induced by waves. The effective waves in this context are generated by easterly and north-easterly winds blowing down the lake. Longshore transport is towards the head of the lake. The beach on the Lake Ontario side of the barrier remains in a natural condition, but the inner shoreline has been considerably altered by landfill. The beach is gravelly at the southern end at Van Wagners Beach, where groynes have been emplaced, but becomes more sandy northwards, towards the Burlington Canal.

Hamilton Harbour, which is contained and protected by the barrier, is bounded by bluffs eroded in surficial materials on the northwest side, but the Hamilton shoreline on the south must originally have been low-lying and fretted with marshy creeks. Figure 2.7 shows the progression of landfill which has completely changed this shoreline in the last hundred years. Four phases of development are indicated. The earliest, prior to 1915, saw the infilling of the heads of the original indentations in the shoreline, a process which was continued in the second phase, 1915-41, which also saw the extension of the shoreline out into the harbour. The northwest corner of the harbour, by Burlington, was also infilled in this second phase. The most active phase of devleopment was the third, 1941-65. In fact much of the activity was carried out in the late 1950s, encouraged by the development of the St Lawrence–Great Lakes Seaway. Systematic development of the harbour took place at this time and eight of ten piers were already constructed by 1965. Canada Centre for Inland Waters was also built on filled land adjacent to the Burlington Canal. Since 1965 the steel companies have completed extensive infilling of their water lots and

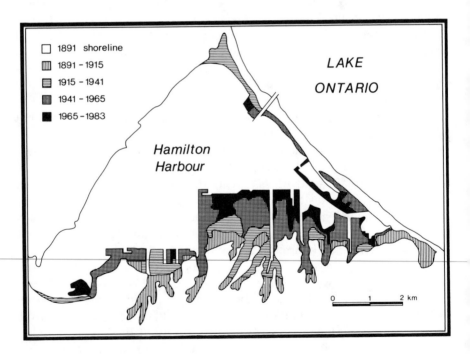

Figure 2.7 Stages in the infilling of Hamilton Harbour

all dredged material has been dumped in the Contained Disposal Facility adjacent to the beach strip, rather than in Lake Ontario.

POST-GLACIAL DRAINAGE PATTERNS AND MODERN STREAMS

The land area draining into the head of Lake Ontario via Cootes Paradise and Hamilton Harbour, shown in figure 2.8, is small and there are no major rivers. The southwest corner of the area shown in figure 2.8 drains southwestwards down the regional slope towards the Grand River and thus into Lake Erie, and the northern strip of the area shown drains eastwards and then southwards via Bronte Creek to the north shore of Lake Ontario. The drainage pattern of the region is immature. The topography above the escarpment is relatively flat, in the broad sense, but it is characterized by a local relief caused by glacial deposition which imposes constraints on drainage development. The drainage network above

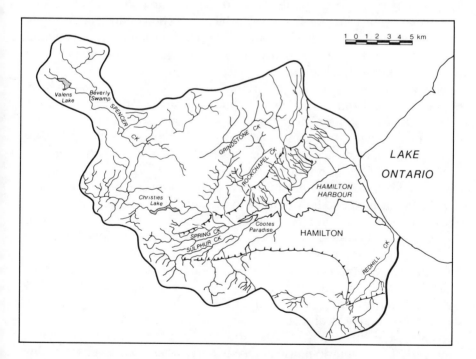

Figure 2.8 Drainage pattern of the catchments which flow into the head of Lake Ontario via Cootes Paradise and Hamilton Harbour

the escarpment exhibits numerous changes of direction, some circuitous drainage routes, and very low drainage divides. There are three types of streams within the combined drainage area shown in figure 2.8: 1 / larger streams, such as Spencer Creek, Grindstone Creek, and Redhill Creek, which have most of their catchment above the escarpment and a pronounced waterfall-gorge section on their long profile; 2 / streams, such as Sulphur Creek and Spring Creek, which drain the drift-filled Dundas Valley and occupy entrenched 'ravine-like' courses; and 3 / the short, relatively steep, streams which rise below the escarpment on the north side of Cootes Paradise and Hamilton Harbour and have entrenched upper courses cut into the Queenston shale and lower courses cut into Lake Iroquois sediments.

Spencer Creek is the largest stream in the Hamilton region with a catchment of 165 km^2 (103 mi^2) and a mean annual discharge of only 1.81 m^3 s^{-1} (64 cfs). It is regulated by two dams at Valens and Christies conservation areas and passes through an area of impeded drainage known as Beverly Swamp. From its headwaters in the Galt moraine, at above 300 m (984 ft) asl, it falls only 64m (210 ft) in the first 30 km (19 mi) of its course as far as Christies Lake. This long, flat section of the creek exhibits many meandering reaches. Overbank flooding in spring is a common occurrence. In the 5 km (3 mi) of the stream course below Christies Dam there is a fall in level of 110 m (360 ft), including Websters Falls, to the point where the creek emerges from Spencer Gorge below the railway viaduct at the head of the town of Dundas. The river then falls another 50 m (164 ft) to enter Cootes Paradise. The section through the built-up area of Dundas is artificially contained. Grindstone Creek and Redhill Creek, on opposite sides of the Dundas re-entrant valley, though smaller than Spencer Creek, exhibit similar features. The moraines of the last ice readvance, the Waterdown moraines and the moraines near Mount Hope, cause the drainage to flow parallel to the escarpment edge. This effect is particularly evident on the north side, where Grindstone Creek and two other smaller creeks, Logie Creek and Rock Chapel Creek, flow parallel to each other and the escarpment.

The deeply entrenched stream courses of Sulphur Creek and Spring Creek provide the essential character of the Hamilton Region Conservation Authority areas in the Dundas Valley: steep-sided ravines, narrow interfluves, and a local relief of 30 to 40 m (100 to 130 ft). These stream courses were cut intially after the drainage of Lake Iroquois and during the early Lake Ontario phase of lake history, when base level was many tens of metres below the present. The lowest parts of the valleys, which are now narrow floodplains, have been built up as lake level rose to its modern level from this low stand. The effects of this phase of lowered base level and valley erosion are evident along the lower reaches of many of the creeks around the head of Lake Ontario.

The hydrologic regime of the modern streams is illustrated in figures 2.9 and 2.10(a) and 2.10(b) by reference to the flow patterns of Sulphur Creek. As with most Canadian streams the annual hydrograph (figure 2.9) is dominated by high flows in spring supplied by melting snow. March and April are the months of highest flow. This period of high flow is made up of a number of snowmelt runoffs, generated in episodes of midwinter as well as spring thaw. The summer, June through September, is the period of low flow because of low precipitation and high evaporation rates. The late autumn and early winter have increased flows as a result of greater precipitation. Also evident from figure 2.9 is the high variability of flow during the spring-snowmelt period compared with the summer period. The mean maximum monthly discharge in March is almost three times the mean monthly discharge for the month. In the period of record shown on the graph (1972-83), mean monthly discharge for March varied from just over 1 m^{3}s^{-1} to just over 14 m^{3}s^{-1} (35-494 cfs). The mean monthly values drawn in figure 2.9 indicate the general trends but mask the extreme events which frequently cause flood and storm erosion drainage.

One of the most intense snowmelt floods on record, which occurred on

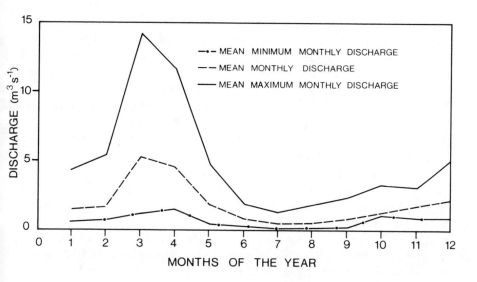

Figure 2.9 Mean monthly discharges on Sulphur Creek for the period 1972-83

11 February 1965, is illustrated in figure 2.10(a). Prior to the event discharge was less than 0.5 m^3s^{-1} (18 cfs) from 1 to 7 February. Warming temperatures and slight rainfall had caused this discharge to increase to 2.0 m^3s^{-1} (71 cfs) on 9 February, but the real trigger to the flood was 2.6 cm (1.02 in) of rainfall and temperatures in excess of 10°C (50°F) on 10 February, when discharge increased to 17.0 m^3s^{-1} (600 cfs). The peak of 24.3 m^3s^{-1} (858 cfs) came the following day and was followed by declining flows, with one exception, for the remainder of the month. Similar floods, though of lesser magnitude, can be expected anytime between January and April each year. Floods also occur in summertime as a response to excessive rainfall, usually as a result of thunderstorms. If the ground surface is baked hard, by a preceding period of hot, dry weather, infiltration is inhibited and surface runoff is very high as a proportion of total precipitation. One such event, which occurred on 11 June 1967, is illustrated in figure 2.10(b). Exceptionally heavy rainfall of 4.75 cm (1.8 in) occurred on 10 June and the response was a mean daily discharge of 11.95 m^3s^{-1} (422 cfs) on the following day. High runoff rates in summer rainstorms commonly cause intense gully erosion on arable land, and the suspended sediment load of silt and clay in local streams may remain high for several days.

THE PHYSICAL LANDSCAPE AND HUMAN OCCUPATION

Two aspects of the physical landscape of the Hamilton region have significantly affected its human occupation. The first concerns the advantages and disadvantages of the physical landscape as a base for settlement and industry and for the development of a major city. The second concerns the landscape as a recreational or cultural resource.

Several features of the physical landscape, described above, have had important consequences in the evolution of the pattern of human occupancy. Chief among these are the protected natural harbour and the restrictive mountain barrier. Hamilton's main industries are based on bulky raw materials which must be transported to the site, and the existence of a natural harbour has facilitated this transport. The presence of flat, low-lying land, with natural creeks, along the southern shore of the enclosed bay led to the early development of harbour facilities there and made easy the later landfill process by which new land for industrial development has been reclaimed from the harbour. The steep bluff shoreline along the northern side of Hamilton Harbour obstructed access to the waterfront, and the dissected shale slopes were unsuitable for devel-

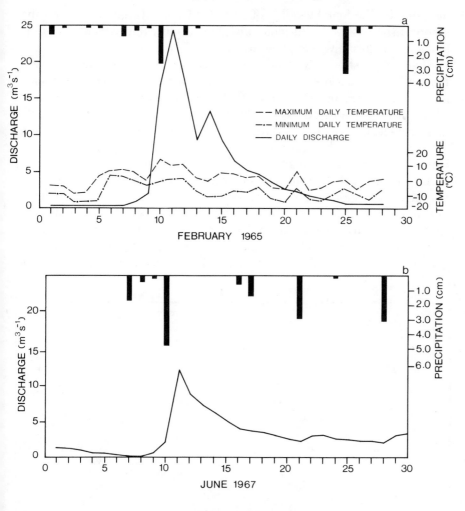

Figure 2.10
(a) Snowmelt flood on Sulphur Creek in February 1965
(b) Rainstorm flood on Sulphur Creek in June 1967

opment. Thus, the harbour and the city developed on the south side of
the bay, somewhat off the direct line of communication between Toronto
and Windsor. The mountain barrier prevented the extension of the city
southwards for nearly a hundred years. Until 1914 the expansion of Hamilton was almost entirely eastwards below the mountain on the shore
plain developed by glacial Lake Iroquois. Expansion to the west was

inhibited by the marshy valley of Chedoke Creek. Today, even with the elaborate access roads, the mountain remains more than a mental barrier, particularly during winter snowstorms.

Figure 2.11 shows the lands managed and owned by the Hamilton Region Conservation Authority and the Royal Botanical Gardens, which total 2,428 ha (6,000 acres) and 809 ha (2,000 acres), respectively, and which provide public access to a wide variety of physical environments and ecological sites. Also shown are the large parks maintained by the City of Hamilton. The Conservation Authority has management and educational functions as well as a mandate to develop recreational facilities, such as hiking and ski trails and opportunities for swimming, fishing, and boating. The region is very well served in this regard. Conservation lands are mainly located in the Dundas Valley, but there are two areas surrounding the artificial lakes on Spencer Creek and the Authority also manages Confederation Park, a Lake Ontario waterfront park in the east end of the city. The Royal Botanical Gardens maintains a number of

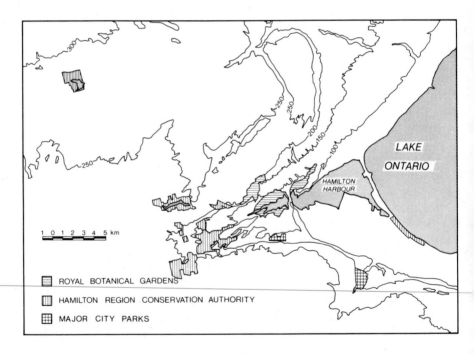

Figure 2.11 Lands owned and managed by the Hamilton Region Conservation Authority and the Royal Botanical Gardens

specialty gardens of international repute and owns much of the land sur-rounding Cootes Paradise, where there is a good network of trails. It also owns escarpment property at Rock Chapel with fine views over the Dundas Valley and eastwards towards Hamilton. The combination of conservancy lands and the properties of the Royal Botanical Gardens ensures the preservation of, and access to, some natural countryside which invokes images of what the landscape was like before the beginning of European settlement.

ACKNOWLEDGMENTS

Drew Hyatt compiled some of the data and maps used in this essay. Personnel of the Hamilton Region Conservation Authority and the Hamilton Harbour Commission were helpful in providing him with information.

REFERENCES

Bolton, T.E. 1957. *Silurian Stratigraphy and Palaeontology of the Niagara Escparment in Ontario*. Geological Survey of Canada, Memoir 289

Karrow, P.F. 1963. *Pleistocene Geology of the Hamilton-Galt Area*. Ontario Department of Mines, Geological Report No. 16 (with four maps, nos. 2029, 2030, 2033, 2034)

Sly, P.G., and Prior, J.W. 1984. Late glacial and postglacial geology in the Lake Ontario basin. *Canadian Journal of Earth Sciences* 21: 802-21

Straw, A. 1968a. Late Pleistocene glacial erosion along the Niagara Escarpment of Southern Ontario. *Geological Society of America Bulletin* 79: 889-910

– 1968b. A geomorphological appraisal of the deglaciation of an area between Hamilton and Guelph, Southern Ontario. *Canadian Geographer* 12: 135-43

CHAPTER 3

Climate, weather, and society

W. R. ROUSE AND A. F. BURGHARDT

SPECIAL FEATURES OF HAMILTON'S CLIMATE (W.R.R.)

For the most part Hamilton's climate is similar to that of other parts of Southern Ontario. However, there are several special aspects which are evident from its temperature, precipitation, and wind patterns. Eight stations have measured temperature and precipitation in Hamilton and Mount Hope during the last 120 years. Of these, six are still in operation. Four stations currently measure wind direction and wind speed.

Table 3.1 presents the data that represent the average climate in the lower city. Several interesting features emerge from the data. The city experienced episodes of intense winter cold and intense summer heat in the latter part of the nineteenth century. Such extremes have rarely been experienced since then. Although the precipitation is remarkably even from month to month, the extremes are notable. For example, more than 12 cm (4.7 in) of precipitation have occurred in one day in both summer and winter.

Going beyond the bare statistics, the city has several features which are important to the urban residents. These are directly related to the proximity of Lake Ontario and Hamilton Harbour, to the Niagara escarpment which splits the city into upper and lower sections, and to the fact that Hamilton is a major industrial city.

Lake Ontario rarely freezes over in winter whereas Hamilton Harbour usually freezes by mid-February, sometimes earlier, sometimes later, depending on the year. Both bodies of water serve as a cold source in spring and summer and a potential moderating source in winter. As such they can and do create strong temperature and pressure changes across the urban coast which can dominate the surface wind patterns. If the lake is colder than the city the wind will tend to blow onshore; if warmer, offshore.

TABLE 3.1
Average and extreme monthly temperature and precipitation at Hamilton (Main Street West), based on the 1866-1958 data period (after Fahrang, 1982)

	Temperature (°C)							Precipitation (mm)		
	Mean daily	Mean min.	Extreme min.	Year	Mean max.	Extreme max.	Year	Ave.	Max. in 24 hr	Year
January	−4.8	−9.0	−30.6	1844	−0.6	18.3	1950	64.8	129.5	1876
February	−4.8	−9.2	−29.4	1881	−0.4	17.2	1954	59.7	69.3	1874
March	−0.2	−4.7	−27.4	1868	4.1	27.2	1946	68.2	139.4	1876
April	6.6	1.4	−14.4	1923	11.7	31.1	1915	67.9	55.9	1886
May	12.7	6.9	− 7.2	1867	18.5	36.1	1911	73.2	80.8	1953
June	18.6	12.6	0.0	1866	24.6	38.9	1870	66.3	88.9	1869
July	21.9	15.9	1.1	1882	27.9	41.1	1868	80.0	126.7	1876
August	20.8	14.9	1.7	1887	26.6	38.9	1948	63.2	94.5	1913
September	16.4	10.8	− 6.7	1873	21.9	37.8	1953	66.3	93.7	1869
October	10.4	5.3	−11.1	1887	15.5	32.2	1951	64.1	92.7	1954
November	3.6	−0.2	−22.8	1871	7.5	25.6	1950	60.6	92.2	1876
December	−2.2	−5.8	−27.8	1871	1.2	16.7	1941	62.3	101.6	1923
YEAR	8.3	3.2			13.2			795.7		

In central Hamilton, winds from the northwest, northeast, and east are largely onshore and winds from the southeast, south, southwest, and west are offshore. In summer, one expects an onshore lake breeze by day when the city is warmer than the water and an offshore land breeze at night if the city becomes colder than the lake. This pattern is illustrated in figure 3.1.

The extent to which the land-sea breeze is operating in Hamilton is given in table 3.2 and illustrated in figure 3.2, which compares wind data from the Hamilton Marine Police Station (HMP) on Hamilton Harbour at the foot of James Street North to data at Mount Hope Airport. Mount Hope shows no day-to-night difference in onshore and offshore wind directions in any season and almost no day-to-night difference between winter and summer. At the HMP station, there are differences in all seasons. In winter, there is 10 per cent more lake breeze by day than by night. In summer, the difference increases to 25 per cent. The winter-summer difference is totally accounted for by a strong increase in the frequency of daytime onshore lake breezes in summer.

One effect of the Niagara escarpment is to funnel the wind. As described by Fahrang (1982), when winds out of the north and northeast encounter the escarpment they are turned clockwise to blow into the Dundas Valley.

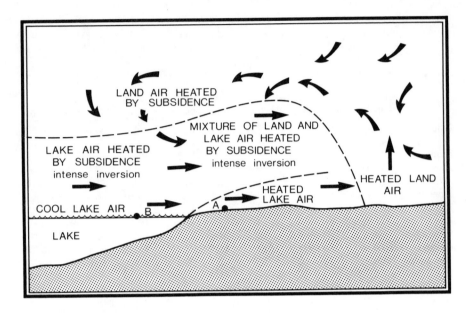

Figure 3.1 Land and lake breezes

The escarpment plays another climatic role. On clear nights the upper city cools more than the lower city. As a result, fog is more frequent in the upper city. Often the cold air, which is denser than the warm air, rolls down the escarpment slopes to create cold-air pockets at the base of the escarpment, producing fog patches in the lower city such as those frequently encountered in southwest Hamilton in spring and autumn. Another important thermal effect of the escarpment is evident in spring and early summer. With light onshore winds, cold air dominates the lower

TABLE 3.2

Onshore and offshore wind frequencies at the Marine Police Station and Mount Hope Airport for day and night periods during winter and summer. Values are in percentages. Winter is October to March, inclusive, and summer is April to September. Day is 0900 to 1800 h; night is 0000 to 0600 h. Adapted from Fahrang (1982)

Station		Day		Night	
		Onshore	Offshore	Onshore	Offshore
Marine Police	Winter	44.0	55.8	33.9	66.1
	Summer	57.7	42.2	33.4	66.1
Mount Hope	Winter	35.0	62.4	34.1	62.8
	Summer	38.5	59.5	37.9	56.4

* Frequencies total less than 100 per cent because of calm periods.

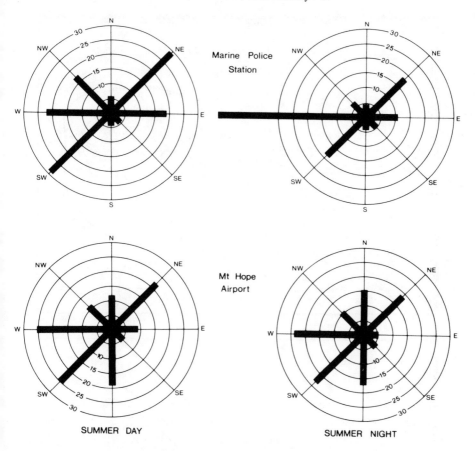

Figure 3.2 Urban and rural wind speed and direction

city and spreads into the Dundas Valley. This cold air is overlain by the warmer air of the regional air masses at about the height of the escarpment. At the division between colder and warmer air there is a sharp temperature inversion (i.e., temperature increases with height rather than decreasing). Under extreme examples a car ascending the escarpment on one of the access roads will suddenly be covered in 'heavy dew' near the top of the escarpment, as the cold car condenses the moisture from the overlying warm, moist atmosphere. Such inversions prevent pollutants from the industrial core from rising into the warm-air layer. They are locked into the lower city and create a pollution dome which is quite visible. Under these conditions the pollutants spread westward up the Dundas Valley. This type of pattern is evident in figure 3.3 from a study

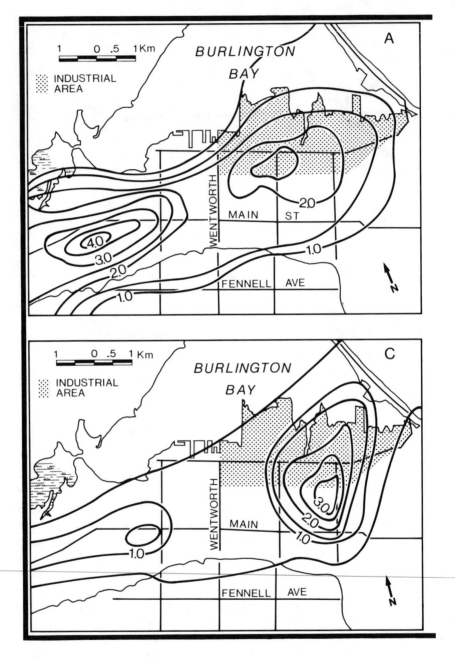

Figure 3.3 Airborne particulate pollution in Hamilton (COH)

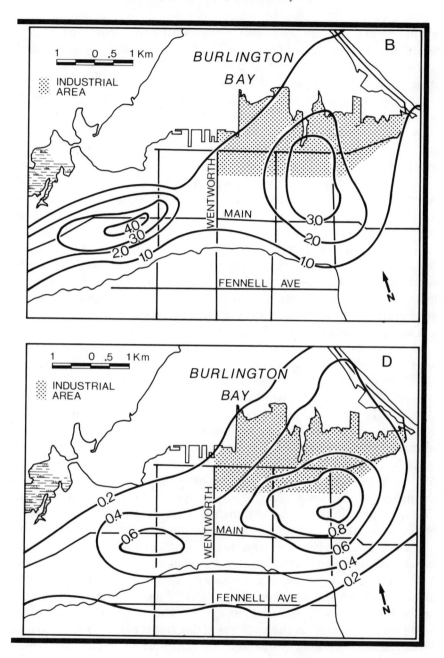

A: SE, E, NE winds; B: N, NW winds; C: W winds; D: S, SW winds

by Rouse and McCutcheon (1970). With east and north winds (A, B), pollutants are concentrated in the central business district and particularly towards the west end of the city. With a change to west winds (C), the pollutants are quickly flushed out of the lower city. With south winds, there is some spillover of the pollution dome into the upper city near the mountain brow, but it is quite minor (D).

When comparing average temperatures at Mount Hope with those in the city it is evident that the city is substantially warmer. Near the central mountain brow, mean annual temperatures are about 1°C (1.8°F) warmer than at Mount Hope, and in the lower city, they are 2°C (3.6°F) warmer. Approximately 1°C (1.8°F) of the warming in the lower city is due to the lower elevation and about 1°C (1.8°F) is due to the urban heat-island effect. Fahrang (1982) notes that in mid-winter the mean wind speed at Mount Hope is 21 km/hr (13 mi/hr) whereas in the lower city it is about 16 km/hr (10 mi/hr). Combining the lower temperatures with the smaller wind speeds gives a wind-chill factor which is much smaller in the city. In terms of human thermal comfort, this indicates a more pleasant environment in winter, a more stressful one in summer.

Oke and Hannell (1970) documented the spatial nature of the urban heat island. They found that the temperature difference within the urban area on a cold winter night could exceed 9°C (16.2°F). The heat island in Hamilton often consists of two warm cells. The larger concentration lies within the industrial zone while the smaller is found in the central business district. Oke and Hannell attributed the industrial cell to heat generated in the continuous steel-making operations, whereas the downtown cell is attributable to nighttime release of heat stored during the day in the asphalt, concrete, and brick fabric of the city core as well as to central heating in the winter season. The height of the heat island is greater in the industrial zone than near the city periphery and often coincides with the top of the pollution dome as shown by Rouse et al. (1972). The presence of the heat island in summer increases the temperature and pressure difference between lake and land thus enhancing the daytime lake breeze but decreasing the nocturnal land breeze. Since nights are short, this has a general effect of increasing the onshore winds in the central city which, in turn, reduces the intensity of the heat island and moves pollutants upward and outward from the city centre. The magnitude of the urban heat island is related to city size and in this respect Hamilton behaves in similar fashion to other North American cities as shown by Oke (1973). Thus the maximum annual urban-rural temperature difference is about

9.5°C (17°F), which compares to that of Vancouver, Winnipeg, and Montreal at 10.0°, 11.5°, and 12°C (18°, 20.7°, and 21.6°F), respectively.

CLIMATIC SEASONS OF HAMILTON (A.F.B)

Climatic statistics are almost always given in averages, which eliminate the temporal accidentals of the weather and allow for both a long-term view and accurate comparisons among localities. Unfortunately, these sanitized statistics serve to eliminate the weather details and the internal sequences which constitute the local climate as it is perceived by the people. Lost in those averages is the actual march of the seasons to which the people fit the details of their lifestyle, their activities, and their clothing.

The use of mean figures suggests, even if it does not state, that our experience is one of a gradual increase in temperature from mid-winter to mid-summer, and then an equally gradual decrease. It implies, too, that we can indeed divide the year into four seasons of equal length with imperceptible transitions between them. In fact, in Hamilton at least, changes can be dramatic: summer and winter are long, springs and autumns shorter, and the seasons tend to arrive suddenly and conspicuously.

Means (averages) also tend to falsify the practical significance of local weather conditions. It does a farmer little good to know that the average last date for a killing frost, as measured several feet above the ground in a favoured location, is in mid-April, when there may be a serious danger of ground frost until the end of May. The average monthly temperatures for April and May do not provide information as to when seed can or should be sown or tobacco and tomatoes transplanted. As figure 3.4 indicates, there was an official frost as late as 18 May in 1983, and serious ground frost on 31 May in 1984.

For more than twenty years, the author has been noting the annual march of maximum and minimum daily temperatures. It is apparent that our climate can be subdivided into seasonal and semi-seasonal segments. Despite the swings in temperature caused by the passage of high-pressure cells, one can discern quantum changes in temperature which separate the seasons. Further, it is evident that the details of the local style of life are carefully atuned to these shifts up and down. Periodically, the weather regime moves to a new level, to a new temperature axis, around which the weather systems oscillate until a new higher or lower axis is established. It is not the intention of this essay to attempt to explain these shifts (the abruptness of which a prominent climatologist has attributed to changes

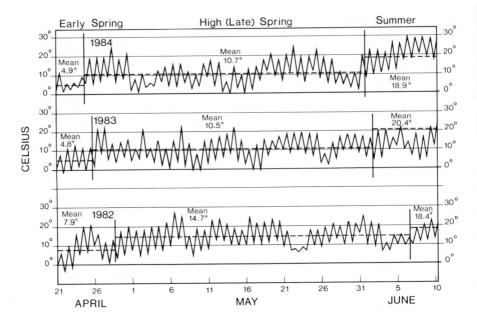

Figure 3.4 Spring frosts

in the path of the jet stream), but rather to note and document their existence, and to point out some of their cultural impacts.

The onset of winter is depicted in figure 3.5. Although the date on which this has occurred has varied from year to year, the existence of the shift seems apparent. Rather suddenly, at the start of December, the temperature axis drops to a level at or below freezing. The median date (the middle date of those noted over all the observed years) for the onset of winter has been 30 November. The change in lifestyle which accompanies the shift is accentuated by the approach of the holiday season. All outdoor work ceases; the ground freezes. Raincoats give way to overcoats, fur coats, and heavy jackets. Young people begin to wear toques and mittens. Even if the weather turns warm for a few days, as it often does, people continue to wear warm clothing and stay indoors.

Winter lasts for about four months (the median length has been 118 days). The final month has a character of its own, but the first three months are impossible to subdivide, though mild spells (i.e., with temperatures above freezing) are more prone to occur in December than in January or February. However, thaws are liable to occur at any time. Conversely, the coldest temperatures can happen in any month; within the

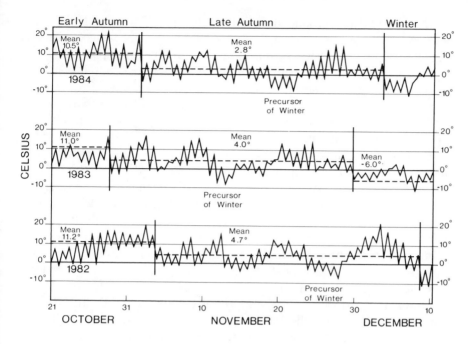

Figure 3.5 Autumn frosts

past decade, the winter minimum has occurred in every one of the four months, including March.

The weather is felt to be cold by most people when the minimum is -10°C (14°F) or lower, and 'brutally cold' at around -20°C (slightly below 0°F). The number of cold days has varied greatly from year to year, with forty-six at -10°C (14°F) or under but only four at -20°C (-4°F) or under being the median figures for Hamilton. In recent years, the number of days for the -10°C (14°F) figure has varied from a total of only twenty in 1982-3 to seventy in 1977-8, while for the -20°C (-4°F) figure it has varied from no days in 1982-3 to eleven days in 1978-9.[1] The coldest recent temperature is -27.8°C (-18°F), recorded on 18 January 1976.

Around the beginning of March, the temperature regime makes its first move upward to establish a distinct sub-season. The temperature axis remains near freezing; hence, this period is still considered to be a part of winter. Certainly in common perception, winter remains as heavy clothing continues to be worn, people complain loudly of the length of the winter, and the great migration to the south reaches its peak.

This transitional sub-season has here been labelled the 'melt-off.' The ice breaks and the ground thaws. It could be called as well the 'mud season,' or more romantically 'maple syrup time.' The skiing season has ended abruptly in Southern Ontario and children begin to play outdoors.

Spring arrives in late March or early April, with 28 March being the median date. The season lasts just two months (sixty-five days), and is thus the shortest of the seasons. It can be subdivided into early spring and high (or late) spring, the two being separated by a perceptible shift upward on the temperature axis.

Early spring extends from around 28 March to 1 May, inclusive. The temperature axis is now above freezing but still well below 10°C (50°F). Light frosts are still common. Outdoor activity begins with the raking of dead leaves and other winter debris. Plants and lawns begin to grow; crocuses and daffodils bloom. The coincidence of Easter with the start of early spring abets the change from winter to 'spring' clothing, a change from dark to brighter colours. Snow tires are removed; cars are 'tuned up.' Hockey, although persisting, seems out of season; baseball arrives.

High spring lasts from around 2 May to 1 June. In early May, there is a notable upward shift in temperature, as can be seen in figure 3.4. This sub-season is probably the most romanticized portion of the year; in fact, this is what people normally mean by 'spring.' The leaves finally open on the trees; tulips bloom, dandelions cover the open spaces. Daylight Saving Time opens evenings to outdoor activities. After seven months of frustration, gardeners seem to go wild with activity, and the insufferable roar of the power mower once again saturates the air of suburbia. Neighbours meet each other again, after the long winter separation. Most concert and theatre seasons end. On farms this is a time for seeding grains.

It is, however, a dangerous time for gardeners. The sun seems warm but the warm air masses are still cool, producing wide variations in temperature between day and night, and the frequent danger of local frosts. 'Don't plant your tomatoes before Victoria Day (24 May)' is probably the most famous of the local adages dealing with the weather. As figure 3.4 shows, both frosts and temperatures in the 20s (70s Fahrenheit) are common.

Summer begins with another upward shift of the temperature axis, and has lasted just one hundred days on the average, from about 1 June through 7 September. As is true of winter, it is impossible to subdivide summer, except for the obvious fact that the longer nights in August and September allow for the easier cooling off of homes in heat waves. There is no definable sub-season in its last month, as there is with winter. However, there is a marked timing of the hottest temperatures; they usually occur during the third week of July.

People tend to feel that the day 'is a hot one' when the official maximum for the city reaches 28°C (82°F).[2] The number of such days has varied widely in recent years, from eighteen in 1981 to forty-five in 1983. The hottest temperature officially recorded recently at Mount Hope Airport was 34.4°C (94°F) on 16 July 1983.[3]

The arrival of summer brings with it a number of social and sartorial changes, although some of these are anticipated in high spring. Outdoor living on patios with barbecues becomes common and the smell of briquettes burning outdoors is the seasonal equivalent of winter's smell of logs burning indoors. Pools are opened and the screams of splashing children are heard throughout suburbia. Soccer leagues commence and churches hold their summer picnics. The owners of cottages begin anew their weekly transmigration between town and country. Repairs are made to houses and cottages.

Like spring, autumn divides into two halves, labelled simply early and late autumn. The total length is about eighty-three days, from around 8 September until 29 November. Frosts, although rare, now become a possibility, and the growers of tender fruits, vegetables, and tobacco generally make certain that they have completed their harvests before the middle of September.

Early autumn is the sub-season which seems to embody all of the central characteristics which most people associate with autumn. The foliage is brilliant. The sun is no longer hot but the air masses are still warm. It is the prime time for hiking. Home owners rake their leaves, clean out the eavestroughs, get their logs in, and generally prepare the outside of the house for winter. The final harvesting takes place.

The temperature axis rests now close to 11°C (52°F). It is a little too cool for short-sleeved shirts, although some people are reluctant to put them away; ties and jackets return. Schools and universities open up during the final week of summer, but the recommencement of formalized education seems to be in close concordance with the arrival of early autumn.

Late autumn can be as depressing as early autumn can be glorious. The world looks grey and brown. Trees are bare and will not again bear leaves for six months. This sub-season begins around 30 October; its onset can be seen in figure 3.5. It is fitting that it begins right around Hallowe'en, when the clocks are set back and night arrives an hour earlier.

The temperature is now below 5°C (41°F). Clouds are low and the microclimatic distinctions between the areas above and below the escarpment become notable. With the ground cooling caused by the long nights, fogs become common. Atmospheric pollution levels reach their annual maxima. Any final outdoor work is now completed. Towards the end of late autumn, the Grey Cup Game is played and a short outbreak of cold air warns of the impending arrival of winter.

A few final observations should be made about the quality of the weather. Many residents seem to have an image of a 'typical' Canadian sky: one that is clear and deeply blue. The author has kept a record of subjective evaluations of the quality of each day. About one day out of five (21 per cent) has proved to be 'perfect,' and in accord with the calendar photo-image of a Canadian day. 'Perfect' is defined as a day with clear blue sky with some billowy clouds, a bright sun in clear air, a day which is not excessively hot or cold, and which is either calm or has a slight breeze. On such days one can easily see the Toronto skyline from the crest of the escarpment over Hamilton. Such days occur in all seasons of the year, with no marked concentration in any season, although they can be relatively rare in November and December. Days with measurable amounts of precipitation make up another quarter of the days, but the proportion often seems higher because of the tendency of rainy days to be clustered.

Despite the fact that there must certainly have been an enormous increase in the amount of carbon dioxide in the air within the past quarter-century, the data show no hint of a 'greenhouse effect.' We have continued to have the occasional severely cold winter, as in 1962-3, 1977-8, and 1983-4. Also, no recent summer has been able to match the continuous heat of 1959, with its fifty-four days over 29°C (85°F) and twenty-four days over 32°C (90°F).

NOTES

1 Only the figures since 1976 are used here. In that year the official Hamilton weather station was moved from the Royal Botanical Gardens, almost on the lakeshore, to Mount Hope Airport above the Niagara escarpment.

2 An official 28°C (82°F) above the escarpment and well south of the city usually means a value of close to 30°C (86°F) in the lower city. Statistics in a 1982 study by A.C. Farhang showed a difference of over 1°C in the maximum values above and below the escarpment.

3 Again, I have limited myself here to the years since 1976. Much higher temperatures were recorded while the station was at the Royal Botanical Gardens. Officially 38°C (100°F) occurred on 2 September 1953, and on both 2 July and 1 August 1955; 37°C (98°F) was reached on 24 August 1959.

REFERENCES

Fahrang, A.C. 1982. *Meteorology and Climatology of Hamilton (Ontario), and Its Surroundings*. Report of Urban Air Environment Group, McMaster University

Oke, T.R. 1973. City size and the urban heat island. *Atmospheric Environment* 7: 769-79

Oke, T.R., and Hannell, F.G. 1970. The form of the urban heat island in Hamilton, Canada. *World Meteorology Organization, Tech. Note*, no. 108: 113-26

Rouse, W.R., and McCutcheon, J. 1970. The effect of the regional wind on air pollution in Hamilton, Ontario. *Canadian Geographer* 14: 271-85

Rouse, W.R., Noad, D., and McCutcheon, J. 1972. Radiation, temperature and atmospheric emissivities in a polluted urban atmosphere at Hamilton, Ontario. *Journal of Applied Meteorology* 12: 798-807

CHAPTER 4

Soils of the Hamilton region: profiles, properties, and problems

BRIAN T. BUNTING

'Soil' can mean different things to different people, for the scientist has no monopoly on the word or on its connotations. Operationally the word usually implies the stuff in which plants grow, or the material that must be moved so that something can be built or buried. However, in some cultures it has unpleasant connotations, of dirt, swine, dung, or worse (if possible); in others it has an honourable connotation of native land or mother earth, the foundation of wealth and of the state.

CONCEPTS OF SOIL

Only thirty years ago, from the scientific point of view, the word 'soil' had very limited scope, despite the very complicated nature of soil as a material. Soil was considered as the basis of farming, a storehouse of water and nutrients by soil physics and agricultural chemistry. The Ontario Soil Survey, when it compiled maps of the soils of the counties of Ontario, relied quite heavily on geological information, for our soils have mainly developed from deposits of the very last or latest glaciation, the Wisconsin, which ended about 12,000 years ago. Most of the physical and chemical properties of our local soils (table 4.1) result from the relatively superficial effects of post-glacial climate in weathering or slightly altering these deposits.

It is possible, too, that other sciences may have had other views of soil, for there have been studies of the biology and microbiology of some of our local soils (Thomson and Davis, 1974) and of their engineering capabilities. Older mid-nineteenth century agricultural reports show that early farmers were as obsessed with measures to control crop pests and diseases, especially in orchards, as with the general advancement of farming and the promotion of rapid transport to the Toronto market and to

TABLE 4.1
Sample physical and chemical properties of some major local soil types

		(1)	(2)	(3)	(4)	(5)		(6)	(7)	(8)	(9)
									Cation		
								Bulk	exchange	Per cent	Per cent
		Depth		Per cent	$\frac{C}{N}$	Per cent		density	capacity	base	iron as
		(cm)	pH	humus		Silt	Clay	(g/cm³)	(meq)	satn.	Fe
Gleysol	A1	0–10	5.2	11.2	9	55	30	1.5	17	57	3.9
(Haldimand)	B2g	40–65	7.4	0.4	5	38	59	1.5	29	97	4.1
	Cg	70+	7.9	–	–	51	43	1.6	27	100	4.6
Luvisol	A1	0–15	5.8	2.5	15	63	4	1.4	13	31	2.9
(Ancaster)	B2t	50–65	6.7	1.2	11	37	25	1.6	13	39	5.0
	IIC	80+	6.5	–	–	42	17	1.8	15	63	4.6
Brunisol	Ap	0–20	5.8	5.2	19	46	13	1.5	16	23	3.9
(Winona)	Bt	25–60	6.0	0.2	6	37	22	1.6	9	57	6.7
	C	95+	8.1	–	–	26	16	1.5	9	94	6.2

Europe before the expansion of Hamilton. Horticulturalists changed the natural soils in the Wellington area quite vigorously in a constructive way (Emery, 1967), adding manures and compost, and eighteenth-century potters sought clay in the Westdale area. Yet their efforts pale in comparison to the soil-alteration activities of rail-, road-, and canal-builders, quarrymen and brickworkers, for in many areas – Aldershot, Dundas, Ancaster, and Stoney Creek especially – one can never be sure that one is looking at a 'natural' soil, away from the remaining wooded or farmed areas, for much is changed, stripped bare of soil, or left as badlands.

With the vast expansion of the Hamilton-Burlington urban area onto farm land, the demand for multiple use of the soil has increased as the result of our apparent need for all-season recreation, for parkland, sod farms, waste disposal (both septic tank and landfill), tree farms, golf courses, as well as for stable roads and housing sites, parking lots and plazas. These needs have diminished the area and volume of topsoil available for water storage and runoff control, and frequently led to more rapid runoff, downstream-flooding, and siltation. Thus, the soil, and our use of it, influence many of our activities, rural and urban, both directly and indirectly.

As with many other things in society, our perceptions of needs may change over time, but just because our urban area has expanded greatly, it does not mean that we can avoid wise use of, or dispassionate research into, the very large areas of open land that are left or, at the very least, try to avoid the vast risks and unnecessary costs of further damage, mis-use, or neglect: for example, some of the soil's properties may be changed rapidly after farming is abandoned; vast areas of weeds, producing air-borne allergens (pollen pollution), occur on unused land; the soil may, or may not, act as a buffer against the effects of acid rain.

THE ORIGIN OF SOILS

Conventionally the appearance or vertical profile of soil, anywhere, is usually integrated as an interaction of *five* factors of soil forma-tion.*Climate*, mainly acting through rainfall, percolation, and temperature variation affecting evaporation, controls the rate of weathering of *rocks and minerals*, be they soft or hard, old or new. This rate is accelerated by input of *organic substances* at the surface derived from natural vege-tation and the humification of leaf and root debris. In addition these effects are cumulative and sequential through *time*, even though our local soils have formed in the relatively brief time since deglaciation. It is important to know whether these weathering reactions are going on in a relatively elevated dry *site*, where the products are mainly oxidative or else mobile and capable of being leached, or whether the area is low lying or water-accumulative, in which case the reactions are seasonally anaerobic, and hence chemically reducing. Such sites also accumulate the water and soluble substances derived from higher sites. A sixth influence, that of *man*, has been more than hinted at – modifying the soils, sometimes constructively, sometimes in drastic destructive fashion.

Most of the mineral soils in the Hamilton area are fine textured or clayey, derived from glacial deposits, in turn derived by glacial action on local or more distant rocks (Karrow, 1963) especially shales of Ordovician age. Though most glacial till deposits are mixtures of many minerals, compacted by thick layers (c. 300 m) of ice, those on the lower land, east or north of the escarpment, share many of the properties of the underlying Queenston shales – the texture of clay or clay loam, the reddish colour, fine mica and feldspar grains, some non-swelling illitic and chlorite clays, as well as some fragments of rock, which came from more distant sources in the Shield to the north (Guillet, 1967).

Fine-textured silty clay loam soils above the escarpment contain much more lime (calcium and magnesium carbonates), up to 30 per cent lime in the total weight of the subsoil, derived from the overriding ice scouring the escarpment dolomites, and these soils are formed on more undulating land in the Vinemount and Waterdown areas, usually much less eroded and dissected by streams than are the lower slopes below the escarpment.

At the other extreme of the soil-texture scale are sands and gravels, derived from glacial outwash and similar agencies – kames and gravel terraces – which are widespread, especially in the Dundas and Ancaster areas and in North Flamborough. Sometimes these sands are thin and form flatland, as in West Flamborough and around Rockton, sometimes thick, gravel-rich, and hummocky as in Freelton, Lynden, and Acton. These lands are prone to drought and have low fertility, and are recognizable by the uncleared cedar, stonewalls, and boulder-strewn fields of northern Wentworth and northwest Halton.

The landscapes around us must have been greatly changing in the immediate post-glacial period, for widespread deposits of very fine-textured, lacustrine silty clays, thin and level, were laid down to the southeast of Hamilton. Similarly many of the sandy deposits of the Grimsby-Winona area and of the Burlington shorelands are deposits of late glacial sand, much re-sorted by shoreline processes at the western margins of Lake Iroquois. More limited areas of fine-textured soils are the varied fine silts of pro-glacial (ice-bordered) lakes as at Sayer Mills and Copetown; the organic soils or peats of Beverly Swamp and the lake mud soils of Dundas harbour; as well as the Aldershot silt (Guillet, 1977), probably of periglacially sorted snowmelt and wind-blown fine materials analogous to those of the present arctic terrain of northernmost Canada (Wilkinson and Bunting, 1975).

SOILS AND LOCAL CLIMATE

Though our local *climate* is far from uniform, varying from place to place and from year to year, its main influence on soil formation (rainfall-evaporation regime) is predictable. The mean annual rainfall varies from about 740 mm to 780 mm (29 in to 31 in) in low elevations (Stoney Creek and Burlington) at 85 m (280 ft) altitude to 840 mm (33 in) at 102 m (335 ft) (Royal Botanical Gardens) and 860 mm (34 in) at 150-200 m (490-650 ft) elevation. The evaporation is about 520 mm (20 in), hence the rainfall surplus, much of it as snowmelt, varies spatially from 200 mm to 340 mm (7.9 in to 13.4 in), and a moisture deficiency occurs only in summer, of at least 100 mm (3.9 in) equivalent for the growing season.

Emerging from snow cover, and a winter freeze which may reach to 50-80 cm (19-31 in) depth, the upper 20-30 cm (8-11 in) of soil is, in April to May, very wet, and is subject to intense leaching and acidification (table 4.1, column 2). The soil in this depth range (0-30 cm [0-11 in]) is usually very different to that below – whether it be on sand or clay deposits – this surface layer having more humus, less clay, and high porosity.

Usually our soils dry fairly rapidly to depths of 50-90 cm (20-35 in) in the hot summer period, with a frost-free period of about 150 days, but the topsoil can be frequently rewetted by summer storms. Some clay soils at the end of summer may be deeply (2 m [6 ft]) and widely (5 cm [2 in]) cracked, especially those with an appreciable content of clay minerals other than kaolinite and illite – the lacustrine clays especially, with a slight content of smectites (expanding clay minerals), which have relatively high amounts of exchangeable calcium and magnesium. The sandy soils dry out intensively to depth, hence their relatively reddish colour and oxidative tendencies. Autumnal rewetting of all soils, before refreezing is accomplished, may lead to leaching of nutrients, loss of nitrates (table 4.1, column 4, large carbon/nitrogen ratios), and dispersal of clay from the upper parts of the soil downwards into the unclosed cracks in the subsoil (table 4.1, B horizons, column 5), this giving a tendency to soil swelling on eventual rewetting, even in soils of low or no smectite content. In many of the flatter areas, underlain by clays, ponding of surface water in the autumn and the spring-thaw periods leads to poor aeration and water-logged soils. Such soils often show the phenomenon of mottling – grey, reddish brown, or orange colour patterns with hard or shot-like iron-manganese concretions – and such features are usually found between 30 cm and 60 cm depth (12-24 in) (figure 4.1, Gleysol). However, they may indicate past poor-drainage states, for trenching, tiling, and subsoiling may have improved matters for current farming, though the diagnostic signs of past soil wetness have long remained.

CHARACTER OF SOILS

In mapping the soils of Wentworth and Halton counties, and the adjacent counties, the Ontario Soil Survey has recognized patterns of distribution of different soil types, named series, distinguished by their relation to different relief sites and drainage states; their varying texture profiles derived from the alteration of the original mineral materials (Ontario Soil Survey, 1963, 1965, 1968). About thirty-five soil series have been identified in Wentworth County and twenty-five of these also occur in Halton, which in turn has ten soil types not found in Wentworth. These basic units have some variation of minor properties, such as surface texture, and they are broadly characterized, by the most important features which they share, in five major soil groups consisting of closely related series (figure 4.1). These groups are Luvisols (previously called Grey-Brown Podzolic soils) which are well-drained clayey soils; Gleysols which are wet, compact,

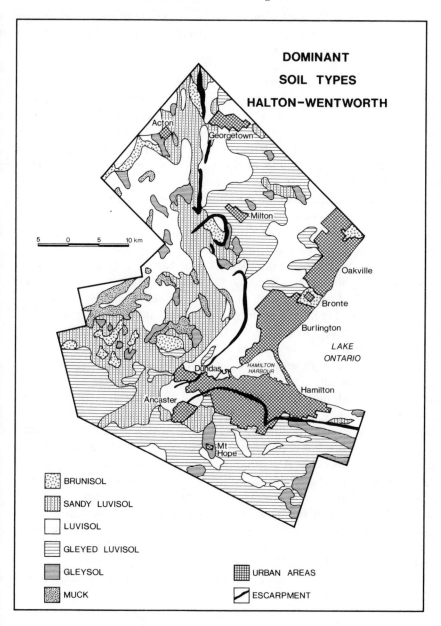

Figure 4.1 Generalized soil distribution patterns in the counties of Halton and Hamilton-Wentworth, compiled from Reports 32 and 43 of the Ontario Soil Survey, Ontario Department of Agriculture, Toronto

mottled, clay soils; organic soils (mucks or peat); Brunisols (formerly called Brown Forest soils), which are usually sandy; and Regosols (or little-developed soils).

The Luvisols: The Good and the Bad

The total combined area of Halton and Wentworth counties is about 212,667 hectares (525,500 acres), and the Luvisols are dominant, covering 143,666 ha (355,000 acres) which is 76 per cent of the total non-urbanized area of 189,400 ha (468,000 acres) or 67.6 per cent of the whole. Most Luvisols are well-drained soils of clay-loam or silt-loam texture, on level land, forming most of our highly productive Class 1 and Class 2 soils, defined as having moderate or no limitations to crop production. Some soils on the older Wentworth till in the north are very stony and often occur on hummocky terrain or are eroded on the steeper slopes. These and some others are largely Class 3 and 4 soils with more severe limitations for intensive agriculture. Most Luvisols have a thin (10-30 cm [4-12 in]) loamy, sometimes fine sandy, surface soil, pale in colour, which is readily leached and eroded, though with enough buffering capacity to resist intense acidification and to respond to fertilizer application (table 4.1, columns 7 and 8, >30 per cent base saturation, moderate cation-exchanging capacity). However, at shallow depths of 15-40 cm (6-16 in), a clayey substrate is present which is very firm (table 4.1, column 6, bulk density >1.5), usually coarsely cracked, with a large reserve of nutrients. This substrate extends to about 80-90 cm (32-35 in) depth before the lime-containing, relatively unaltered, compact glacial till clays are reached (see table 4.1 and figure 4.1) (Bunting and Hathout, 1972). This strong contrast of texture promotes some problems: relative wetness in the thaw period (about 40 per cent of the Luvisols are imperfectly drained); the need for heavy fertilization, especially with nitrogen and phosphorus, to promote high yields, which is then partly leached away before crop roots can reach the more clayey substrate. In addition, compaction by cultivation or heavy traffic is common and slope wash or wind erosion is not uncommon. However, when well managed, fertilized, and limed, very high levels of production can be obtained. The use of many of the better-drained Luvisols is now intensified, with tree or soft-fruit cultivation, horticulture, mono-culture cereals, or vines.

Gleysols: Not All Bad

The Gleysols cover about 26,700 ha (66,000 acres) in Halton-Wentworth, about 12 per cent of the area, and dominate much of Binbrook and south

Saltfleet. Gleysols show signs of drainage impedance, with either permanent or intermittent waterlogging, especially of the soil surface on level areas with a perched watertable even though the subsoil at 1 m (3 ft) may still be dry and hard. This surface wetness may be increased by extensive runoff from higher slopes, as occurs south of Vinemount and around Milton and Tansley.

Most Gleysols are considered to be Class 2 land – usually those which have been cultivated and thoroughly drained – or Class 4 land, where they have not responded to drainage or are subject to frequent and prolonged waterlogging. The mineral horizons of Gleysols are usually greyish, with distinct mottling at 40-60 cm (16-24 in), even at 15-30 cm (6-12 in) in the most poorly drained variants, greyish-brown below these depths and with many concretions of iron-manganese compounds adding variety to the soil colour (figure 4.1 [c]). Often the surface horizons have poor tilth when subject to heavy rainstorms. The unaltered clay is usually very compact and is found at relatively shallow depths, usually not more than 60 cm (24 in) (figure 4.2).

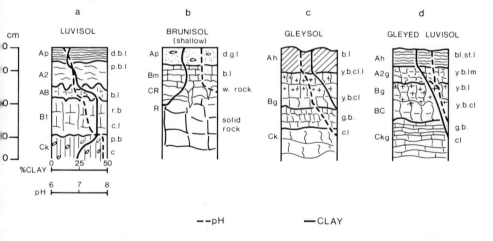

Figure 4.2 Typical profiles of local soils. The vertical distribution of clay content and change of pH of the soil are shown. Abbreviations: d = dark, l = light, y = yellow, b = brown, r = red, g = grey, c = clay, l = loam, s = sand, w = weathered. The soil horizons: A = altered surface soil, B = soil with accumulated weathering products, C = slightly altered geologic material; h = humus-rich, g = gleyed or mottled, k = contains lime, t = clay enriched.

Though these soils have a high base status or nutrient content and high amounts of organic matter relative to Brunisols and Luvisols (table 4.1), they have many unfavourable features for agriculture as well as for recreation, building, and urban use. Aeration is poor, drying is late in the season, rooting depth is limited, and the rate of organic matter decomposition is slow. Often some fertilizer materials, phosphorus especially, are not released adequately, and the weak physical structure and stickiness when wet are further drawbacks to intensive use. Some of the Gleysols are relatively unstable – swelling when wet and heaving under frost, destabilizing ditches, tiles, even tracks and roads – as a result of the presence of smectite clays, or the effect of frost action, or both. Unless well-managed for crops and pasture, the tilth and/or sward can deteriorate quite rapidly. Adjacent parts of Lincoln County leading to the drainage systems of Twenty Mile Creek and the Welland River have similar extensive areas of Gleysols.

Organic Soils

Wet peat, muck, and areas of wet marshland are not extensive, but distinctive, covering about 8,500 ha (21,000 acres) in the two counties. The largest area is Beverly Swamp, with black, friable, mesic, or humic peat – partly or wholly decomposed organic soils – though usually a woody peat occurs at depths of 75 to 90 cm (30-35 in) or more, underlain by thin bands of calcareous marl on the underlying till or limestone bedrock. Many small peat swamps occur in Nassagewaya Township, moderately decomposed shallow (<1 m [3 ft]) deposits, but most are left in a natural state and are wildlife refuges, as is much of Beverly, though its eastern periphery, near Highway 6, is cultivated, excavated, or used for Christmas-tree plantations.

Brunisols

The well-drained Brunisols cover about 10,000 ha (25,000 acres) and their limited distribution somewhat belies their variety, interest, and importance. Most are sandy in texture, sometimes fine sand and loam, but some are quite coarse and gravelly with high infiltration rates, even when wet. The former mainly occur in the lake plain – Burlington, Stoney Creek, and Grimsby – and may be calcareous at about 60 cm (24 m) depth. Relatively fertile with adequate base status or fertility, Brunisols have a moisture regime that is adequate base status or fertility, Brunisols have a moisture regime that is helped by flat sites and a clayey substrate at depths <1 m (3 ft). Those in north and west Wentworth and around

Campbellville and Georgetown are on coarse sandy ice-front deposits, with variable stratigraphy. The minerals present may be quartz, shale, and sandstone fragments, or, occasionally, the gravelly outwash deposits combine limestone and a finer sandy component, usually less than 20 per cent being weathered micas and feldspars (Grubb and Bunting, 1976). Many of the less-coarse Brunisols, as well as some of the deeper, more-leached Luvisols, show some effects of iron mobilization in reddish-tinged Brunisols, the result of more intense weathering (table 4.1, column 9). Most Brunisols are dry soils, with leached slightly acid topsoils, with variable though usually low-nutrient status in the subsoil (table 4.1, column 7). Given good management and fertilizer input, perhaps overhead irrigation, mulching, and manure, the finer sandy soils have, or had, a wide range of uses, being excellent for tillage, for a wide variety of farm, fruit, and vegetable crops, for tender fruits, especially cherries and peaches in the south, which can best be grown on deep, well-drained fine sands. They respond well to intensive fertilizing for vegetables grown for processing and for horticultural multi-cropping. It is a pity that so many areas of Brunisols have been lost to urbanization, though much could still be made of them for vines, garden vegetables, and floriculture.

SOILS AND EXPANDING URBANIZATION: SOME PROBLEMS

The use of soils in an expanding urban area is influenced by many other needs and activities, such as campgrounds, recreation and sports facilities, house foundations, landscaping, waste disposal, septic-tank operation, safety of underground utility lines, roads, sod farms, and gravel extraction. Sometimes, agricultural exploitation of soil materials goes indoors, for tomato, cactus, fruit, and mushroom production; others become independent of soil – such as broilers- and mink-production – though their waste needs to be disposed on a suitable soil or land type. Soil can be regarded as having the ability to process or absorb some pollutant materials or, in being eroded or leached, it can itself become a pollutant.

In our local environment, the soils which presented most problems for agriculture also provide difficulties for non-farm usages. The Gleysols and imperfectly drained Luvisols on compact clays, the wet soils on peats or valley-floor sediments, as well as the thin gravelly or sandy soils overlying bedrock, are far from efficient or reliable, over the longer term, for septic-bed installations, for underground utilities, for surfaced minor roads, and, in many cases, for recreational trails, campgrounds, and playing fields. Seepage, corrosion, misalignments and surface-heaving, and failures of ground cover and gully erosion are all quite common occurrences on the clays; apart from the obvious disadvantages of rockiness, high and exposed elevations, moisture-deficient and steep eroded slopes in the thinner soils near the escarpment, and of periodic flooding, sediment-loaded streams, churned-up surfaces, and drainage problems on the clays and peats.

Soil as a Pollutant Processor

One of the most controversial uses of land is that of waste disposal. In the case of septic beds – and about 10,000 households still have these in place in Halton-Wentworth – most of them, fortunately, have been constructed in moderately stable fine sandy or loam materials overlying clays. What is not often realized is that with time these artificial soil mounds change and deteriorate; porosity is decreased by internal deposition of organics; and the soil fabric of the underlying clay changes through various chemical exchanges induced by continuous leaching, while the capacity of the loamy bed to adsorb phosphates, mineralize ammonia, and process organic components decreases. This effect is noticeable, particularly after wet summers or after intense winters, with freeze limiting the active volume, increasing the intensity and period of use, and increasing the volume of seepage (Campbell, 1974).

Though the Ottawa Street and Aldershot landfill sites obviously spring to mind, with the burial, or at least partial interment, of vast quantities of municipal garbage in slightly less than leak-proof excavated sites, other lesser sources of potential pollution are present: the spreading of manure or animal droppings from intensively stocked horse, pig, mink, poultry, turkey, and feedlot enterprises, not to mention geese and seagulls on the lakeshore, together with the use of waste-water effluents as liquid fertilizer and the crafty disposal of a carload of domestic waste or diseased plants. The survival of organic wastes contained in most of these sources is very much enhanced in moist soils and under low temperatures. Our local counties have 40 per cent of their area in such soils and our climate is characterized by low temperatures. Thus such materials, if applied on dry soils beyond the capacity of sandy soils to filter them, or of clay soils to adsorb them, will inevitably migrate to the streamsides (Campbell, 1974). However, application of treated sewage effluent and sludge to closely managed areas – parks, golf courses, and tilled soils – involves fewer risks.

One of the major problems is of solid-waste disposal – the euphemistically named sanitary landfill sites – and the hope is that while such disposal sites are an environmental problem when under construction and in filling, they will not cause further problems on closure, if properly managed. The ideal is to excavate a vast hole in leak-proof impermeable clays or shales, which should neither receive drainage water from its surroundings nor have lines of surface seepage near to, or issuing from, its margins. The leak-proof substrate can be doubly assured either with plastic liners or imported compacted bentonite clay, the whole inspected for crackless perfection before receiving its initial load. This load is, in turn, covered by bulldozed soil, to prevent wind-blown debris and dust, limit rodent and bird activity, as well as to apply sufficient soil mass to absorb some dissolved leachates from the next application of overlying waste and generally retard percolation within the mass. Eventually it is capped by topsoil and revegetated, creating new land from badland or a former abandoned excavation.

There are many difficulties in this ideal scheme. The needed excavations replace an in situ sterile hardened rock or clayey material with a rapidly drying oxidized periphery, while the slightest possibility for seepage is enhanced by the increase of hydraulic gradient towards the excavation. A plastic liner is not immune to disruption or breakage, and excavated recompacted clays develop abundant microfissures, stress cracks, and slip planes, increasing permeability and the risk of leakage. The placement of wet, organic-rich garbage and natural precipitation undiminished by evapotranspiration gradually increase the reduction processes along cracks and planes and controlled drainage or pumping can significantly alleviate peripheral leaching in the long term, as well as create a disposal problem of surplus water when, and if, the leachate is led to inground sewage, not storm runoff, systems.

At the Aldershot site, initially uncontrolled dumping of domestic refuse on shaley till and the clayey Oneida Luvisols led to massive movement of leachates (figure 4.3), an increase of chloride release from the shales of 10 to 200 times the background level existing in the unaffected soils. In the adjacent formerly unpolluted surface waters, the electrical conductivity was increased to values of 600 μS, exceeding 60,000 μS in leachates in the most contaminated sites. Thus, in any landfill site, initial tests of permeability and prediction of effects can be seriously misleading. The vertical hydraulic conductivity is much lower than the lateral permeability of partially weathered thin sandy layers and of shales with expanding micro-fissures, and any variation can cause channelling or 'plumes' of contaminant solutions. The full pollutant plume at Aldershot moved about 400 m (1,300 ft) in less than twelve months and was detected at 1.2 km (3,900 ft) within six years. Data show that the groundwater in the major plume area is perennially higher, occasionally floods the surface at thaw,

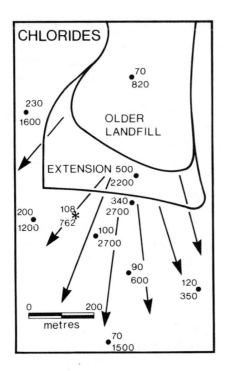
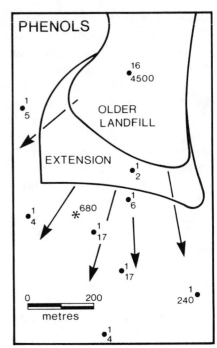

Figure 4.3 Increase of chlorides and migration of phenols adjacent to landfill area, 1972-83, Aldershot, Ontario. Upper figures denote initial year and lower figures, maximum concentrations in the period. The starred site is the toe drain. The background level of chlorides is less than 20 mg l⁻¹; of phenols <7 ppb, usually zero.

and the surface dries more slowly in the summer. The impact of this added leachate has increased chloride release from the weathered shales to more than 100 times background levels, hence the great increase in conductivity (table 4.2). The total basic cations (calcium, magnesium, sodium) released are 150 times greater than the background levels in unpolluted soils, and the input of oxidized nitrates from the ammonia-rich anaerobic landfill into groundwater, where it can easily migrate, is 5 to 12 times background levels at peak periods. Of some concern is the great increase in BOD (biological oxygen demand) of soils, subsoils, and flowing waters and the occasional very high BOD readings in the remedial toe drain, which still functions, flowing towards a sediment trap. Such figures, in such circumstances, indicate that any accidental or erosional breach would release

TABLE 4.2
Some effects of landfill leachate on adjacent soils and drainage lines

	Background (natural)	Source (landfill)	Leachate plume	Marginal site	Polluted rill	Stream	Toe drain*
Chloride (Cl) (mg l⁻¹)	1–20	70–800	40–2700	230–1,600	100–325	20–30	100–750
Conductivity (µS)	250–1,200	2,100–8,800	460–61,000	3,000–7,200	1,300–2,150	500–970	650
Sum of cations $CA+Mg+Na+K$ (mg l⁻¹)	100–300	n.d.	110–15,000	650–1,600	130–180	n.d.	1,400
Sulphate (SO_4) (mg l⁻¹)	<50	40–760	50–2,400	270–1,600	180–430	30–230	280
Nitrate (NO_3) (mg l⁻¹)	<1	<0.1(!)	0.3–7.6	0.3–3.0	1.7–12.3	0.1–5.3	<0.2
Ammonia (mg l⁻¹)	<0.1	2–240	4–17.5	0.4–4.0	1.8–12.3	<0.2	107
BOD (mg l⁻¹)	<20	360–6,400	3–66	2–26	6–585	<5	600–2,200
Phenols (mg l⁻¹)	0–7	16–4,500	<1–21	<1–5	<1–240	<1–5	800

NOTE: The 'geography' of the landfill sites is shown in figure 4.3. Data from Final Technical Reports, Environmental Assessment, Halton County, June 1985

highly charged leachates, and that one may anticipate higher readings in future.

The danger in such situations is that any movement of polyphenols in leachate waters towards areas with such high concentrations of chlorides could lead to the synthesis of halomethanes, particularly in acidifying, reduced soil environments enriched with humus substances leached from the organic substances in the fill and reacting with the chlorine. Looking at table 4.2, one should recall that 20 parts per billion is a limit for polyphenols in most recommendations for a safe environment. Clearly wide buffer zones, completely integrated drainage and leachate control systems, and adequate long-term monitoring are needed in old, abandoned, current, and future landfill sites. Attempting their rehabilitation solely by use of a thin cover of compactable or erodible soil is inadequate. In the absence of large, established energy-from-waste facilities, the greater evil of garbage burial will provide complex problems of management and restorations for years, if not generations, to come (Novotny and Chester, 1981).

Who Needs Topsoil?

A less drastic but potentially deleterious user of land is the sod farm. Granted that the operator has a vested interest in good land management over the short term for the sake of a quality product, and that many building developers, having shipped away and sold good topsoil, have no apparent alternative but to use sod; the constant stripping of sod and topsoil, even 2-3 cm (0.8-1.0 in) at a time, will denude wide areas of porous surface soil which, as we have seen, has better tilth, aeration, and permeability than the compact clayey subsoils. A loss of a 20 cm (8 in) thickness of topsoil, with a density of 1300 kg m^{-3} (80 lb ft^{-3}) over an area of 100 ha (247 acres), for example, involves the loss of 260 million kg (more than a quarter of a million metric or Imperial tons) of topsoil. If we assume that this soil has about 50 per cent porosity, the potential for moisture storage of that area, within a drainage basin, is reduced by at least 65 mm (2.5 in) of rainfall. In other words, that area will fail to store about one-tenth of the annual precipitation, which will run off rather rapidly, and will not be stored for consumptive use in, or slow release from, the original soil cover.

No Praise for Idleness

Some uses of soil, such as many farming and recreational uses, are essentially or ideally conservative, others are locally additive, and yet others are quite clearly extractive or partially destructive. However, leaving the soil alone or unused is not always a wise measure. Such occurs on at least 8,000 ha (20,000 acres) of idle land waiting development – sometimes for periods of 15 to 20 years or more – and such a measure hardly leads to soil improvement. Studies of a site in Stoney Creek (a long-abandoned cherry and pear orchard, with some old vines), over a twelve-year period, show that the pH of the uppermost 20 cm (8 in) of soil has declined from 6.7 to 5.5 in that time (1972-84) and that the organic matter content has increased from 0.6 per cent (of dry weight) or 5 per cent by volume, to 1.9 per cent by weight (15 per cent by volume) (figure 4.4). Though the site is only 6.5 km (4 mi) south-southeast of Parkdale Road, it is hard to decide whether acid rain, industrial fall-out, or increasing organic matter content from weed decay is the cause of increasing acidity. Any future user of such land, be it a private or commercial concern, is faced with fertilizer, insecticide, and pesticide use on a vast scale. As it exists now, such land is a source of wind-borne weed seeds and allergens for the existing urban population and a source of pests, diseases, and rodents for the productive farm lands near to it.

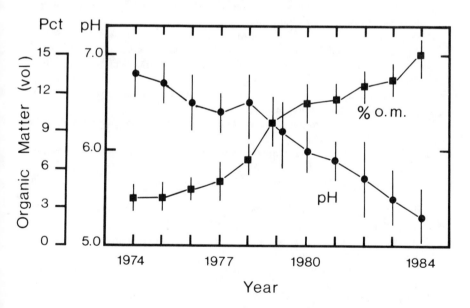

Figure 4.4 Acidification and increase of organic matter in the surface soil of an abandoned orchard, 1974-84. The vertical lines show the variation in samples in each year of measurement.

Many of our local soils are sensitive and highly susceptible to abuse. With careful management and conservation they can become, and on many properties are, highly productive, but this result requires intensive training and capital which, like time, are in increasingly short supply. Many of our local farms are increasingly run on a part-time or rental basis, and long-term management and intensification are not well served on such systems. Many of our local non-farmed soils are in a state of crisis, some because of past abuse, some because of new and expansive forms of less than optimum use.

REFERENCES

Bunting, B.T., and Hathout, S.A. 1972. Relationship between macro- and micro-morphology of certain podzols and luvisols in southern Ontario. *International Geography 1972*, vol. 1: 247-51. University of Toronto Press, Toronto

Campbell, J.A. 1974. Some impacts of septic effluent on the soil in Ancaster and Grindstone Creek watersheds. Unpublished MSc thesis, Department of Geography, McMaster University

Emery, C.D. 1967. *From Pathway to Skyway: A History of Burlington.* Confederation Centennial Committee of Burlington

Grubb, A.M., and Bunting, B.T. 1976. Micromorphological studies of soil tonguing in the Burford loam, southern Ontario. *Biuletyn Peryglacjalny* (Lodz, Poland) 26: 237-52

Guillet, G.R. 1967. *Clay Products Industry of Ontario.* Ontario Department of Mines, Toronto, Report IMR 22

– 1977. *Clay and Shale Deposits of Ontario.* Ontario Geological Survey, Report MDC 15, Ministry of Natural Resources, Toronto

Karrow, P.F. 1963. *Pleistocene Geology of the Hamilton-Galt Area.* Ontario Department of Mines, Geology, Report no. 16

Novotny, V., and Chester, G. 1981. *Handbook of Non-point Pollution.* Van Nostrand Reinhold, New York

Ontario Soil Survey, 1963. *The Soil Survey of Lincoln County.* Report no.34, Canada and Ontario Departments of Agriculture, Ottawa and Toronto

– 1965. *Soils of Wentworth County.* Report no. 32

– 1968. *Soils of Halton County.* Report no. 43

Thomson, A.S., and Davis, D.M. 1974. Mapping methods for studying soil factors and earthworm distribution. *Oikos* 25: 199-203

Wilkinson, T.J., and Bunting, B.T. 1975. Overland transport of sediment by rillwater in a periglacial environment in the Canadian High Arctic. *Geografiska Annaler* (Sweden) 57A(1-2): 102-15

CHAPTER 5

Forests of the Hamilton region: past, present, and future

G.M. MacDONALD

When one thinks of the 'natural vegetation' of the Hamilton region, the image of the forests which flank the Niagara escarpment or lie alongside Cootes Paradise usually comes to mind. Those with a romantic inclination might imagine these scattered woods to be the relics of some great primeval forest which existed unchanged through millennia until the beginning of land clearance by Europeans in the last century. However, like other facets of the natural and cultural landscape of the Hamilton region, the forest vegetation is dynamic and has witnessed periods of change throughout its long history. Since shortly after the commencement of deglaciation 14,000 years ago the vegetation has been dramatically affected by environmental change, epidemics of plant diseases, and the hand of man. Attempts to understand the modern forest vegetation of the Hamilton region must be predicated upon an appreciation of the long and dynamic history of vegetation change. Fortunately, evidence to reconstruct the history of the forests can be drawn from the fossil-pollen record, the archaeological record, and the writings of early explorers and settlers. In this essay I will provide a brief account of the modern forest vegetation of the Hamilton region, outline the history of vegetation change in southwestern Ontario during the last 13,000 years, and make some observations regarding factors which may be important in determining the future of the forest.

MODERN FOREST VEGETATION OF THE HAMILTON REGION

The modern forests of the Hamilton area occupy a zone of transition between two major forest regions (figure 5.1). Vegetation zonation at this large a scale is primarily a result of macro-climatic control. The most important climatic factor controlling the distribution of major vegetation

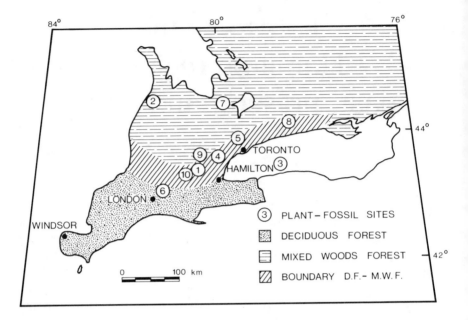

Figure 5.1 The major forest vegetation zones of southwestern Ontario and the location of fossil-plant sites discussed in this chapter: (1) Kitchener, Gage St Bog (Anderson, 1982); (2) Kincardine deposit (Karrow et al., 1975); (3) Lake Ontario pollen core site (McAndrews, 1973); (4) Crawford Lake (McAndrews, 1976); (5) Van Nostrand Lake (McAndrews, 1976); (6) Pond Mills Pond (McAndrews, 1981); (7) Edward Lake (McAndrews, 1981); (8) Rice Lake (McAndrews, 1984); (9) Brampton Bog deposits (Teresmae and Matthews, 1984); (10) Maplehurst Lake (Mott and Farley-Grill, 1978)

zones in Ontario appears to be temperature (McAndrews, 1981). To the south of the Hamilton region lies the hardwood-dominated deciduous forest. To the north lies the mixed conifer-hardwoods Great Lakes–St Lawrence forest (Rowe, 1972). Recent assessments of the forests of Southern Ontario suggest that the forests of Hamilton and vicinity might be considered either a southern extension of the Great Lakes–St Lawrence mixed-woods forest (Moss and Hosking, 1983) or a northern extension of the deciduous forest (Rowe, 1972). However, as we shall see, the forests of Southern Ontario have been changing throughout the last 13,000 years and such classification schemes are really no more than attempts to delineate and describe the most recent configuration of an extremely dynamic vegetation.

The location of Hamilton in a zone of transition between two different forest regions contributes to the existence of a diverse modern tree flora (table 5.1). It is apparent that a number of tree species from the boreal and Great Lakes–St Lawrence forests reach their southern range limits in Canada near Hamilton. Similarly, a number of tree species from the deciduous forest reach their northern limits here. This mixture of northern and southern tree species contributes to the diverse arboreal flora found in the forests surrounding Hamilton.

TABLE 5.1
The common and scientific names of trees found in the Hamilton region

Eastern white pine	*Pinus strobus*
*Red pine	*Pinus resinosa*
*Tamarack	*Larix laricina*
*White spruce	*Picea glauca*
*Black spruce	*Picea mariana*
Eastern hemlock	*Tsuga canadensis*
Balsam fir	*Abies balsamea*
White cedar	*Thuja occidentalis*
Red cedar	*Juniperus virginiana*
Willows	*Salix* spp.
Trembling aspen	*Populus tremuloides*
Large-toothed aspen	*Populus grandidentata*
Balsam poplar	*Populus balsamifera*
Eastern cottonwood	*Populus deltoides*
Butternut	*Juglans cinerea*
†Black walnut	*Juglans nigra*
Shagbark hickory	*Carya ovata*
†Pignut hickory	*Carya glabra*
Bitternut hickory	*Carya cordiformis*
Hop hornbeam	*Ostrya virginiana*
Blue beech	*Carpinus caroliniana*
Yellow birch	*Betula alleghaniensis*
*Paper birch	*Betula papyrifera*
Beech	*Fugus grandifolia*
†Chestnut	*Castanea dentata*
White oak	*Quercus alba*
Bur oak	*Quercus macrocarpa*
Swamp oak	*Quercus bicolor*
Chinquapin oak	*Quercus muehlenbergii*
Red oak	*Quercus rubra*
Black oak	*Quercus velutina*
White elm	*Ulmus americana*
Rock elm	*Ulmus thomasii*
Slippery elm	*Ulmus rubra*
Hackberry	*Celtis occidentalis*
†Red mulberry	*Morus rubra*

TABLE 5.1 (continued)
The common and scientific names of trees found in the Hamilton region

†Tulip tree	*Liriodendron tulipifera*
†Paw Paw	*Asimina triloba*
†Sassafras	*Sassafras albidum*
Witch hazel	*Hamamelis virgininana*
Sycamore	*Platanus occidentalis*
Mountain ash	*Sorbus americana*
Showy mountain ash	*Sorbus decora*
Hawthorns	*Crataegus* spp.
†Wild crab apple	*Malus coronaria*
Cherry and plum	*Prunus* spp.
Sugar maple	*Acer saccharum*
Black maple	*Acer nigrum*
Silver maple	*Acer saccharinum*
Red maple	*Acer rubrum*
Basswood	*Tilia americana*
†Flowering dogwood	*Cornus florida*
Alternative leaved dogwood	*Cornus alternifolia*
White ash	*Fraxinus americana*
Red ash	*Fraxinus pennsylvanica*
Black ash	*Fraxinus nigra*

*Boreal and mixed-woods species which reach their southern range distributions in the Hamilton region.
†Deciduous species which reach their northern range limits in the Hamilton region.

    The local distributions of tree species within forests are controlled by a number of physical and biological factors. Important physical factors include the micro-climate, hydrology, and soil characteristics. Important biological factors include competition from other plants and animal activity. The distributions of trees in the forest may be thought of as continuous curves along physical or biological gradients. Most research on the gradient distribution of plants has concentrated on physical gradients, which are easier to identify and quantitatively analyse. Work by Maycock (1963) demonstrated that the modern forest vegetation of Southern Ontario can be described as tree species distributed along a soil-moisture gradient. Figure 5.2 presents the curves illustrating the importance of major tree species along the soil-moisture gradient, ranging from dry to mesic to wet. A brief description of the local distribution of major tree species in the forests of the Hamilton region based on the work of Maycock and others, as well as my own observations, is provided below.

    Red, black, and white oaks (for scientific names, see table 5.1) tend to dominate on dry upland sites. Other deciduous trees often encountered on dry sites include shagbark hickory, hop hornbeam, sassafras, flowering

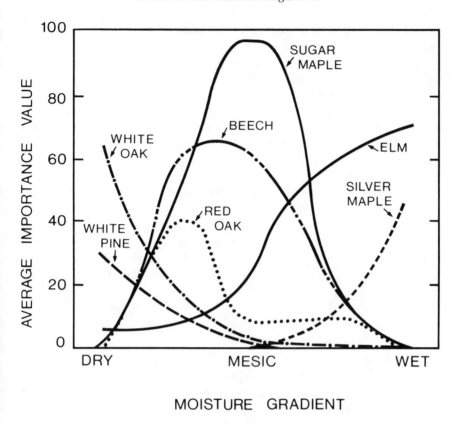

Figure 5.2 The distribution of some important southwestern Ontario tree species along a moisture gradient (after Maycock, 1963)

dogwood, white ash, black walnut, large-toothed aspen, and black cherry. Trembling aspen is found on recently burnt or disturbed dry sites. Chestnut was also common on dry sites prior to its destruction by an introduced pathogen (plant disease). Conifers which may be encountered on dry sites include white pine, red pine, red cedar, and white cedar.

Mesic upland sites are frequently dominated by sugar maple and beech. Other deciduous trees commonly encountered on mesic sites include basswood, black and red maple, bitternut hickory, white ash, butternut, black walnut, tulip-tree, black cherry, hop hornbeam, and blue beech. Yellow birch is often encountered on disturbed mesic sites. White elm was found on mesic sites until its recent decimation by an introduced pathogen. Eastern hemlock is a common conifer encountered on mesic sites. White pine is often found on sites which have been opened up by fire or other disturbance.

Wet lowland sites are typified by white, red, and black ash; red and silver maple; eastern cottonwood; swamp and bur oak; basswood; blue beech; and yellow birch. Wet bottomland sites alongside rivers are often occupied by willow, eastern cottonwood, sycamore, balsam poplar, silver and black maple, white elm, slippery elm, red mulberry, and hackberry. Acid fen and bog sites sometimes contain boreal species such as black spruce and tamarack as well as white cedar and various bottomland deciduous species.

The modern distribution of forested sites in the Hamilton region is presented in figure 5.3. Approximately 20 per cent of the land within a 10 km (6 mi) radius of Hamilton can be considered forested. The largest forest stands are located in stream valleys, wetlands, and along the slopes of the Niagara escarpment. The forest stands on the adjacent Niagara escarpment uplands which border Hamilton to the north and west are small and widely scattered.

POST-GLACIAL HISTORY OF FORESTS IN SOUTHWESTERN ONTARIO

The available evidence suggests that southwestern Ontario was free of glacial ice by approximately 14,000 years ago. A great deal of information on vegetation development from the time of deglaciation to the present can be obtained by the analysis of fossil pollen and plant macrofossils (twigs, leaves, seeds, etc.). Fossils which span the time from shortly after deglaciation to the present day are available from a number of sites in the London-Hamilton-Toronto-Barrie region (figures 5.1 and 5.4). These plant remains were obtained from sediment samples taken from small lakes and bogs. In most cases, their ages were determined by radiocarbon dating. A notable fossil-pollen record of post-glacial vegetation change has also been recovered from sediments in Lake Ontario (McAndrews, 1973). Although radiocarbon dates are not available from the Lake Ontario sediment cores, the fossil-pollen record can be dated approximately by correlation with the pollen records from radiocarbon-dated sediment taken from smaller lakes.

Fossil-pollen grains are particularly useful for the reconstruction of long-term vegetation history. Many important eastern North American tree species are wind pollinated and produce great amounts of pollen during the spring and summer. Much of this pollen settles out on the forest floor,

FOREST   URBAN   OTHER

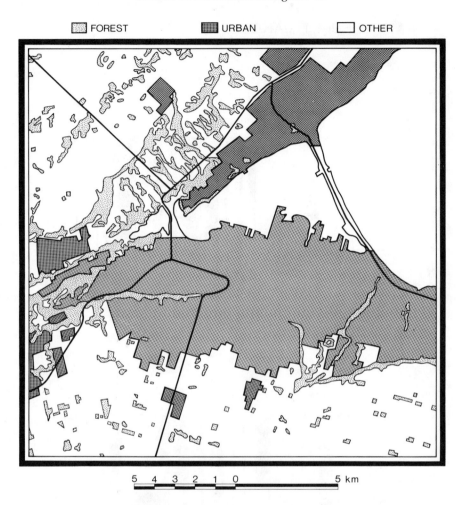

5 4 3 2 1 0   5 km

Figure 5.3 Modern (mid-1970s) distribution of forest in the Hamilton region.
Map was compiled from 1:50,000 Canadian topographic maps and b/w and
infrared airphotographs.

into streams, or onto lake surfaces. Pollen is resistant to decomposition,
and grains which settle out in oxygen-free environments such as lake
bottoms and bogs may be preserved indefinitely. Trained analysts can
usually determine the genus and sometimes the species of fossil-pollen
grains. Fossil-pollen records are generally presented in the form of multiple

MAPLEHURST LAKE

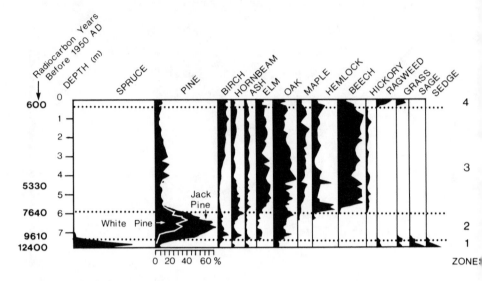

EDWARD LAKE

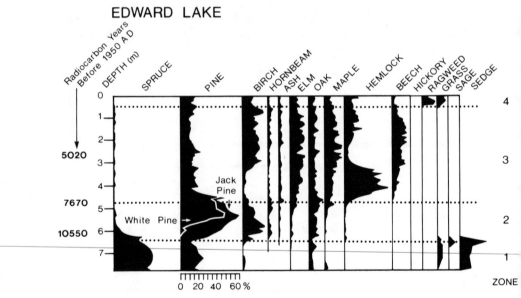

Figure 5.4 Examples of post-glacial fossil-pollen records obtained from southwestern Ontario (see figure 5.1 for locations)

POND MILLS POND

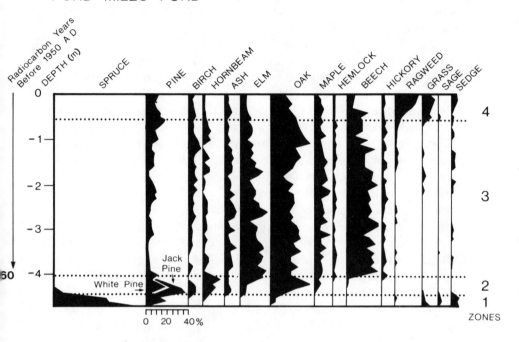

LAKE ONTARIO

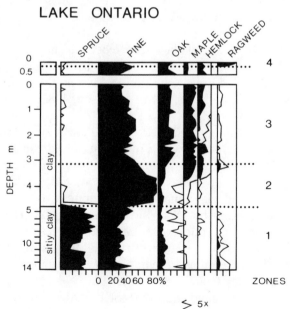

histograms or curves with time or sediment depth serving as the vertical axis and the percentages of different pollen types being portrayed along the horizontal axis (figure 5.4).

The oldest post-glacial fossil pollen and plant macro-fossils available from southwestern Ontario date from approximately 12,500 to 10,500 years ago, and are from Kincardine, Kitchener, Brampton, London, and Barrie (figures 5.1 and 5.4). However, it is possible that even older post-glacial plant fossils exist and remain to be discovered. The 12,500-year-old fossil assemblages contain appreciable quantities of spruce, sedge (Cyperaceae), sage (*Artemisia*), and willow, reflecting a forest-tundra vegetation roughly similar to that found today in the Hudson Bay–James Bay region of Northern Ontario (McAndrews, 1981; Anderson, 1982; Teresmae and Matthews, 1984). However, surprisingly high frequencies of fossil pollen from temperate tree species such as oak, maple, ash, and pine are often recorded. The presence of these temperate types may be a result of long-distance transport of pollen from forests growing to the south of Ontario or the redeposition of much older fossil pollen eroded from melting glacial ice and till (McAndrews, 1981). Early post-glacial plant macro-fossil assemblages also often include the needles and cones of spruce, the needles of tamarack, and the twigs and leaves of white cedar in association with the leaves and seeds of tundra and forest-tundra herbs and shrubs such as arctic avens (*Dryas drummondii, Dryas integrifolia*), least willow (*Salix herbacea*), and tundra sedges (*Carex bigelowi* and *Carex aquatilis*) (Anderson, 1982; Teresmae and Matthews, 1984). The fossil evidence indicates that both white and black spruce became increasingly important in the vegetation of southwestern Ontario as the forest component of the forest-tundra expanded during the period from approximately 12,000 to 10,000 years ago (McAndrews, 1981; Anderson, 1982; Mott and Farley-Gill, 1978).

McAndrews (1981) has suggested that the early forest-tundra vegetation of southwestern Ontario was a reflection of cold climatic conditions. He estimated that mean annual temperatures may have been 15°C (27°F) colder during the early post-glacial than they are today. These results are similar to early post-glacial temperature estimates provided for the western Great Lakes region by Webb and Bryson (1972). However, other factors such as slow migration rates for certain tree species and the disturbed soil conditions of the early post-glacial may have also had a significant impact on determining the nature of the vegetation cover at this time (Mott and Farley-Gill, 1978; Davis, 1981).

Fossil-pollen evidence indicates that the boreal forest species jack pine

(*Pinus banksiana*) very rapidly became important in the vegetation of southwestern Ontario approximately 10,000 years ago (figure 5.4). The fossil evidence also suggests that other trees, including fir, birch, oak, elm, maple, and ash, appeared in the vegetation of southwestern Ontario at this time. The vegetation of this period is generally interpreted as a relatively dense mixed conifer-deciduous forest. Jack pine decreased in importance during the period from approximately 9,000 to 8,000 years ago. This decline in jack pine was contemporaneous with a dramatic increase in white pine.

Approximately 8,000 to 7,000 years ago white pine decreased in importance in the vegetation while trees such as hemlock, beech, elm, and maple became increasingly important components of the forest (figure 5.4). The decrease in white pine was more acute in the region occupied by the modern deciduous forest than in the modern mixed-woods Great Lakes–St Lawrence forest region (figures 5.1 and 5.4). This episode may be interpreted as representing the establishment in the south of a deciduous forest dominated by beech and maple and a mixed-woods forest with a significant amount of pine in the north.

The transition from spruce-dominated forest to closed mixed woods and deciduous forests is a result of an increase in mean annual temperature (McAndrews, 1981; Karrow et al., 1975). However, the impact of other factors such as differences in the migration rates of different tree species into Ontario from the south and progressive changes in soil conditions cannot be discounted.

Fossil-pollen evidence indicates that approximately 5,000 to 4,000 years ago a significant change took place in the deciduous and mixed-woods forests of eastern North America. At that time one of the important tree species, eastern hemlock, almost completely disappeared from the forest vegetation (figure 5.4). The temporal and spatial pattern of its decline suggests the impact of an extremely virulent plant disease (Davis, 1981). The sites which were previously occupied by hemlock were probably taken over by associated deciduous trees such as birch, beech, and maple (McAndrews, 1984). Hemlock must have developed a resistance to the unknown pathogen by between 4,000 and 3,000 years ago. At that time hemlock populations increased back to near their pre-decline importance in the vegetation (Davis, 1982).

The next series of major changes in the forest vegetation of southwestern Ontario appears to reflect the direct impact of man. Humans were present in Southern Ontario shortly after deglaciation. However, it was not until after the introduction of Indian agriculture to Ontario approximately 1,500 years ago (Trigger, 1976) that the human population had an impact on the vegetation which is clearly detectable in the fossil-pollen record. The Iroquois were the Indian group residing in the Hamilton region from the introduction of Indian agriculture until the early seventeenth century when intertribal warfare and disease introduced by European fur traders decimated the Indian populations. Maize, beans, and squash appear to have been the most important crops for the Iroquois. The Indians practised a slash-and-burn style of agriculture (Trigger, 1976). Fields were cleared of trees by cutting and burning. Crops were then grown on the open land for several years until the soil had lost its initial fertility. It is likely that the Iroquois had marked preferences in the forest types which they cleared for farming. Mesic sites with maple, oak, basswood, beech, hemlock, and white pine were apparently favoured. Poorly drained sites with cedar, birch, and other swamp species were avoided (Konrad, 1975). Accounts by Champlain suggest that Iroquois villages had populations numbering up to the thousands. The villages would be continuously occupied for periods of eight to twenty years until all of the surrounding land had been exhausted by slash-and-burn agriculture (Trigger, 1976; Konrad, 1975).

An exceptional fossil-pollen record of the impact of Indian agriculture on vegetation is available from Crawford Lake near Milton (figures 5.1 and 5.5). The sediment of the lake contains annual laminations called

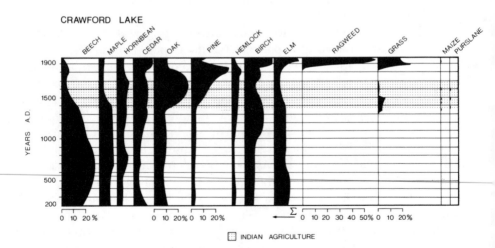

Figure 5.5 Summary fossil-pollen record from Crawford Lake
(see figure 5.1 for locations)

varves which allow palaeobotanists to examine the fossil pollen and plant macro-fossils on a year-by-year basis for the last thousand years (McAndrews, 1976). The analysis of fossil pollen from the varved sediment indicated that a beech-maple forest was predominant in the Crawford Lake region prior to 1360. Pine and oak appear to have become increasingly important after this date. A detailed study of the Crawford Lake fossil-pollen record produced evidence of Indian agricultural activity during the period of pine and oak increase. Most notably, the evidence includes the presence of maize pollen and purslane (*Portulaca oleracea*) pollen (a weed and an Indian pot-herb) in sediment deposited during the period from 1360 to 1650 (Bryne and McAndrews, 1975). Archaeological excavations around Crawford Lake unearthed the remains of an Iroquois village of approximately 500 people which was inhabited during this period (Bryne and McAndrews, 1975). McAndrews (1976) suggested that the slash-and-burn agriculture practices of the Iroquois led to widespread disturbance of the forest which allowed populations of fast-growing shade-intolerant trees such as white pine, oak, and poplar to expand at the expense of longer-lived shade-tolerant species such as beech and maple. There is evidence of a similar pattern of increasing oak and/or pine dominance in the forests of southwestern Ontario during the period of Iroquois agriculture near Toronto, Peterborough, and London (figures 5.1, 5.4, and 5.5). The fossil-pollen evidence suggests that by 1650 Indian farming ceased to be important in the Crawford Lake region and throughout southwestern Ontario as Iroquois populations largely disappeared. However, an even more potent form of forest disturbance was shortly to be introduced to southwestern Ontario – European land use.

European hunters, trappers, and missionaries visited southwestern Ontario frequently from the mid-seventeenth century onwards. However, it was not until after the American War of Independence in 1784 that large-scale settlement of Ontario by Europeans began (Gentilcore and Wood, 1975). This allowed for a lag of from 150 to 200 years between the demise of widespread Iroquois farming and the initiation of widespread land clearance and farming by Europeans. Information on the forest vegetation which the first European settlers encountered in southwestern Ontario can be gleaned from historical documents such as the notes of explorers, settlers, agriculturalists, etc. (Moss and Hosking, 1983; Maycock, 1963). For example, Gourlay (1822) records these observations regarding Southern Ontario: 'In 1784 the whole country was one continued forest. Some plains on the borders of Lake Erie, at the head of Lake Ontario, and in a few other places, were thinly wooded ... The forest trees most common are beech, maple, birch, elm, bass, ash, oak, pine, hickory, butternut, balsam, hazel, hemlock, cherry, cedar, cypress, fir, poplar, sycamore … whitewood, willow, spruce.' A floristic analysis of early-nineteenth-century forestry data compiled by Gourlay suggests that the Halton–Wentworth–Niagara Peninsula region was dominated by pine-oak-ash forests at the time of first settlement by Europeans (Moss and Hosking, 1983). Bowman (1980) compared the distribution of pine stands as recorded in an 1797 Royal Navy forest survey to the distribution of prehistoric Indian village sites and found a close correspondence. This suggests that the pines were numerous in the late-seventeenth- and eighteenth-century vegetation of southwestern Ontario because the fast-growing shade-intolerant pines were able to take advantage of and grow on the abandoned Indian fields (McAndrews, 1976). Thus, the 'primeval' forests which were encountered by the early European settlers actually reflected the hand of man in the form of Indian agricultural clearance and field abandonment.

The reference by Gourlay (1822) to open plains areas in the Hamilton region at the head of Lake Ontario is interesting and is certainly at odds with the romantic picture of a dense and seemingly endless forest in southwestern Ontario prior to European settlement. Examination of early survey notes from 1816-17 indicates that in the Brantford region a considerable proportion of the land (~30 per cent) was occupied by open oak-dominated woodland and treeless meadows (Wood, 1961). The explanation for these oak plains is not clear. One possibility is that the open plains were a result of repeated burning by the Indians to clear and open up the land (Wood, 1961). However, it is worth noting that much of the former oak-plains regions in the Hamilton area are generally underlain by dry sandy soil. It is possible that soil conditions rather than Indian burning were responsible for the oak plains.

During the 1780s, approximately 9,000 United Empire Loyalists settled in Ontario. Although the European settlers cultivated maize, the major crop was wheat (Gentilcore and Wood, 1975). The early European settlers had definite preferences regarding sites to be cleared for wheat cultivation (Kelly, 1975). Settlers felt that sites which were forested by a mixture of deciduous trees were prime locations for wheat. Pine-dominated sites were considered to indicate poor soils for wheat cultivation. Wet areas with cedar and tamarack were avoided by settlers because of drainage problems and heavy soils. It was thought that the oak-plain areas were attractive because they were easy to clear and were well drained, but wheat yields were likely to decrease after a few years' cultivation. Agricultural land clearance proceeded rapidly and most well-drained upland sites were cleared for farming in the first half of the nineteenth century. The drainage and clearance of wet lowland sites proceeded rapidly during the 1870s and 1880s (Kelly, 1975).

The forests were also dramatically exploited for timber. By the second quarter of the nineteenth century, the value of timber exports from Ontario exceeded that of agricultural products (Gentilcore and Wood, 1975). Pine was the major timber tree harvested and exported from the north shore of Lake Ontario and fires were common in cut-over regions leading to the further loss of trees (Head, 1975). Hamilton served as a major port for the shipping of timber from southwestern Ontario, and as late as 1871 significant footage of pine was harvested in the Wentworth-Halton region (Canada Census of 1871).

Figure 5.6 presents the distribution of forested regions which remained in the Hamilton area at the beginning of this century. The most marked change in the last seventy years has been further shrinkage and fragmentation of forest lots on the Niagara escarpment uplands (figure 5.2).

The impact of European land-clearance on the forest vegetation in southwestern Ontario is also clearly recorded in fossil-pollen assemblages (figures 5.4 and 5.5). The frequency for tree pollen decreased dramatically while pollen of the weedy plants such as grass and ragweed, which took advantage of the newly cleared landscape, increased to unprecedented high frequencies.

FUTURE FOREST-VEGETATION CHANGE

In the previous sections of this essay, I have described the modern forests of the Hamilton region and outlined the history of forest vegetation in southwestern Ontario. It is now appropriate to consider factors which may lead to future changes in the forests of the Hamilton region. Many natural and anthropogenic mechanisms may have a dramatic impact on the forests even if no further clearance and destruction of forest land for agriculture and urban expansion takes place. For illustrative purposes, I will discuss three major factors which are operating today to shape the future forests of the Hamilton region. These three factors are plant disease, acid rain, and the population dynamics of the tree species themselves.

It is probable that the decline of hemlock in the forests of Ontario 5,000 years ago was the result of the rapid spread of an extremely virulent disease. In this century, three important tree species have been affected by rapidly spreading and virulent pathogens which were introduced to North America from abroad.

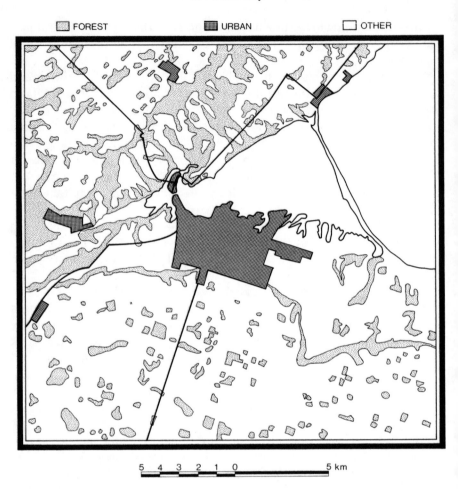

Figure 5.6 Distribution of forest in the Hamilton region at the beginning of this century. Map was compiled from 1 inch: 1 mile Canadian topographic maps.

Chestnut trees have been almost completely driven to extinction throughout North America by the fungus *Endothia parasitica* (Elliston, 1981), which kills the trees by girdling the trunk. However, lower portions of the infected chestnut often remain viable and produce sprouts. Unfortunately, these sprouts serve as repositories for the fungus and are usually killed before they are old enough to produce seeds. The fungus was introduced to North America from Asia in 1904. Prior to the spread of the disease, chestnut trees made up as much as 25 per cent of the standing timber in the eastern deciduous forest. The loss of chestnut trees is producing major changes in the forest vegetation. The dry and mesic sites previously occupied by chestnut have been invaded by other tree species. For example, a recent study in the southeastern United States concluded that red oak and white oak and hickory have replaced chestnut in the forest vegetation (McCormick and Platt, 1980). Attempts to halt the decimation of chestnut by chemical means and breeding programs have not proved effective. Some hope for stopping the destruction of the trees may lie in the production and dissemination of a strain of the chestnut-blight fungus which is less virulent and will replace the present North American strain (Elliston, 1981).

White elm has also been subjected to decimation by a fungal pathogen (*Ceratocystis ulmi*). The fungus was introduced from Europe during the early part of this century and had spread throughout eastern North America by 1970. The mortality of elms in infected dense stands on wet lowland sites reaches to between 95 per cent and 100 per cent (Barnes, 1974). However, the mortality of the more scattered upland populations is considerably less. Prior to infection, white elm was common on mesic sites and was often dominant on wet lowland sites in the northern deciduous forest. In Michigan white elm may be close to extinction on wet lowland sites and is being replaced by swamp oak and maple (Richardson, 1976). However, the high reproductive capacity, high dispersal rate, and lower stand density of white elm on upland sites will allow the species to persist. Unfortunately, white elm will be much less important, frequent, and long lived than it was prior to the spread of the disease.

A third major species, beech, is currently undergoing widespread decline as a result of the impact of a bark fungus (*Nectria* spp.). The disease was first noted in Nova Scotia at the turn of the century and has since spread throughout eastern North American. The degree of mortality to beech trees appears to vary greatly with soil conditions, stand densities, and the species of the other trees growing in association with beech. Yellow birch and sugar maple may take advantage of sites cleared of beech by the disease. However, hemlock appears to benefit most from the removal of beech. Although the impact of the disease is severe, it does not appear that beech will be eradicated completely as an important tree species by the infection (Twery and Patterson, 1984).

Acid rain, precipitation which contains high amounts of sulphuric and nitric acid derived from atmospheric sulphur and nitric oxides produced by the burning of coal and other fossil fuels, poses a major risk for forest vegetation in the highly industrialized Hamilton region. Recent studies show that forests which lie adjacent to industrial centres in western and eastern Europe are experiencing massive deterioration as a result of acid rain. Similar widespread destruction of forest vegetation as a result of acid rain has now been reported in several studies from the northeastern United States (Siccama, et al., 1981; Wetstone and Foster, 1983). Severe acid-rain conditions may lead to direct damage of plant tissue and reduced plant growth. However, the indirect effects resulting from acid-rain–induced changes in forest soils appear to be more important (Siccama, et al., 1981; Wetstone and Foster, 1983). Acid rain can strongly affect soil conditions by leaching nutrients, releasing metals which are toxic to plants, and inhibiting the activity of important soil micro-organisms. Severe leaching by acid rain may potentially produce nutrient-depleted soil which will not be able to support forest vegetation of any form (Wetstone and Foster, 1983). Some of the damage from acid rain may be repaired by extensive liming of the forest soils. However, this is at best a temporary measure and is impossible to carry out on a very large scale. Of course, the best cure for acid rain is a reduction in sulphur-rich emissions by power plants and industry.

Even if phenomena such as plant disease, air pollution, climatic change, and further clearance of forest lands were stopped, dramatic changes in the remaining forests of the Hamilton region might be brought about by the biological interactions of the trees themselves. Weaver and Kellman (1981) recently examined the arboreal flora of small wood lots in Southern Ontario. They recorded whether tree-species populations were increasing, remaining stable, or declining in each of the sites. They found that shade-intolerant species such as white pine were becoming extinct in many of the lots. After extensive analysis of factors such as forest-lot size and topographic diversity, Weaver and Kellman concluded that shade-intolerant tree species were declining because natural disturbances such as fire were being excluded from the lots. Shade-intolerant trees were being shaded out by long-lived shade-tolerant tree species. They suggested that high tree-species diversity might be maintained in the remaining forests of Southern Ontario through the use of artificial-disturbance regimes such as controlled forest burning.

The above examples illustrate that, like the other natural and cultural components of the geography of the Hamilton region, the forest vegetation remains dynamic and prone to both dramatic and subtle changes. The impact of plant disease and competition between shade-tolerant and shade-intolerant tree species may lead to a significant decline in the diversity of the forest vegetation. Acid rain could lead to the destruction of the forest vegetation altogether. Much of the future changes in the forests of the Hamilton region may be controlled directly or indirectly by man. The remaining forests in the Hamilton region are the legacy of more than 13,000 years of past vegetation change and development. The forest heritage that we will leave for the future remains an important concern.

ACKNOWLEDGMENTS

I thank Drs K.D. Bennett, R.L. Gentilcore, and J.H. McAndrews for providing comments and suggestion on an earlier draft of this chapter.

REFERENCES

Anderson, T.W. 1982. Pollen and plant macrofossil analyses on late Quaternary sediments at Kitchener, Ontario. *Geological Survey of Canada Paper* 82-1A: 131-6

Barnes, B.V. 1974. *Succession in American Elm Communities.* University of Michigan Rackman Report

Bowman, I. 1980. The Draper Site; white pine succession on an abandoned late Prehistoric Iroquois maize field. In Melvin (1980): 109-37

Byrne, R. and McAndrews, J.H. 1975. Pre-Columbian purslane (*Portulaca oleracea* L.) in the New World. *Nature* 253: 726-7

Davis, M.B, 1981. Quaternary history and stability of forest communities. In West, Shugart, and Botkin (1981): 132-53

Davis, M.B. 1982. Outbreaks of forest pathogens. *Proceedings of the IV International Polynological Conference*, Lucknow, 3: 216-27

Elliston, J.E. 1981. Hypovirulence and chestnut blight research: fighting disease with disease. *Journal of Forestry* 79: 657-60

Gentilcore, R.L., and Wood, J.D. 1975. A military colony in a wilderness: the Upper Canada frontier. In Wood (1975): 32-50

Gourlay, R. 1822. *Statistical Account of Upper Canada with a View to a Grand System of Emigration.* Simpkin and Marshall, London

Head, C.G. 1975. An introduction to forest exploitation in nineteenth century Ontario. In Wood (1975): 78-116

Karrow, P.F., Anderson, T.W., Clarke, A.H., Delorme, L.D., and Sreenivasa, M.R. 1975. Stratigraphy, paleontology, and age of Lake Algonquin sediments in southwestern Ontario, Canada. *Quaternary Research* 9:48-97

Kelly, K. 1975. The impact of nineteenth century agricultural settlement on the land. In Wood (1975): 64-77

Konrad, V.A. 1975. Distribution, site and morphology of prehistorical settlements in the Toronto areas. In Wood (1975): 6-31

McAndrews, J.H. 1973. Pollen analysis of the Great Lakes of North America. *Proceedings of the III International Polynological Conference*, Moscow: 13-15

– 1976. Fossil history of man's impact on the Canadian flora: an example from southern Ontario. *Canadian Botanical Association Bulletin Supplement* 9: 1-6

– 1981. Late Quaternary climate of Ontario: temperature trends from the fossil pollen record. In Mahaney (1981): 319-33

– 1984. *Late Quaternary Vegetation History of Rice Lake, Ontario, and the McIntyre Archaeological Site*. National Museum of Man Mercury Series, 126: 161-89

Mahaney, W.C., ed. 1981. *Quaternary Paleoclimate*. Geo Abstracts, Norwich

Maycock, P.E. 1963. The phytosociology of the deciduous forests of extreme Southern Ontario. *Canadian Journal of Botany* 41: 379-438

McCormick, J.F., and Platt, R.B. 1980. Recovery of an Appalachian forest following the chestnut blight: Catherine Keever – you were right! *American Midland Naturalist* 104: 264-73

Melvin, D.S., ed. 1980. *Three Heritage Studies on the History of the HBC Michipicoten Post and on the Archaeology of the North Pickering Area*, Ontario Ministry of Culture and Recreation, Toronto

Moss, R.M., and Hosking, P.L. 1983. Forest associations in extreme Southern Ontario ca 1817: a biogeographical analysis of Gourlay's *Statistical Account*. *Canadian Geographer* 27: 184-93

Mott, R.J., and Farley-Gill, L.D. 1978. A Late-Quaternary pollen profile from Woodstock, Ontario. *Canadian Journal of Earth Sciences* 15:1101-11

Richardson, C.J. 1976. An analysis of elm (*Ulmus americana*) mortality in second-growth hardwood forest in southeastern Michigan. *Canadian Journal of Botany* 54: 1120-5

Rowe, J.S. 1972. *Forest Regions of Canada*. Canadian Forestry Service, Ottawa

Siccama, T.E., Bliss, M., and Vogelmann, H.W. 1981. *Decline of Red Spruce (Picea rubens Sarg.) in the Green Mountains of Vermont*. University of Vermont, Burlington

Teresmae, J., and Matthews, Hobo. 1984. Late Wisconsin white spruce (*Picea glauca* [Moerch] Voss) at Brampton, Ontario. *Canadian Journal of Earth Sciences* 17: 1087-95

Trigger, B.G. 1976. *The Children of Aataentsic: A History of the Huron People to 1660*. McGill-Queen's University Press, Montreal

Twery, M.J., and Patterson, W.A. 1984. Variations in beech bark disease and its effects on species composition and structure of northern hardwood stand in central New England. *Canadian Journal of Forestry* 14: 565-74

Weaver, M., and Kellman, M. 1981. The effects of forest fragmentation on woodlot tree biotas in Southern Ontario. *Journal of Biogeography* 8: 199-210

Webb, T., and Bryson, R.A. 1972. Late and postglacial climatic changes in the northern midwest U.S.A.: quantitative estimates derived from fossil pollen spectra by multivariate statistical analysis. *Quaternary Research* 2: 10-115

West, D.C.F., Shugart, H.H., and Botkin, D.B., eds. 1981. *Forest Succession: Concepts and Application*. Springer-Verlag, New York

Wetstone, G.S., and Foster, S.A. 1983. Acid precipitation: what is it doing to our forests? *Environment* 25: 10-12, 38-40

Wood, J.D. 1961. The woodland–oak plains transition zone in the settlement of western Upper Canada. *Canadian Geographer* 5: 43-7

– ed. 1975. *Perspectives in Landscape and Settlement in Nineteenth Century Ontario*. McClelland and Stewart, Toronto

CHAPTER 6

Hydrology of Beverly Swamp

MING-KO WOO

Hydrology studies the occurrence, distribution, and circulation of water at or near the terrestrial surface of our planet. Through such studies, we gain an understanding of the physical processes which govern the movement and storage of water, thus allowing a sensible management of the water resource and avoiding unnecessary disruption of the natural hydrological environment. This essay provides a case study of a simple hydrological system to reveal its basic processes and to examine the effects of man-made disturbances on its natural rhythms.

STUDY AREA

The Hamilton region is drained by several rivers, one of which is Spencer Creek. This stream has been regulated to control floods and to provide recreational opportunities for the residents of the region. One area traversed by Spencer Creek is Beverly Swamp which offers a wetland habitat for waterfowl, fish, deer, and other animals. Such an ecological setting enhances recreational activities, ranging from bird watching to sport fishing and the hunting of game animals.

As a wetland, Beverly Swamp has a water table that is very close to, or lies above the ground surface for a protracted period of each year to favour the formation of soils that support the growth of water-loving plants. This swamp is located 25 km (16 mi) northwest of Hamilton, and covers an area of 20 km² (12 mi²), with an elevation ranging between 265 m (869 ft) and 270 m (886 ft) above sea level. Geologically, it occupies two depressions in the dolomite bedrock which were infilled with clastic sediments, from gravel to clay, up to a maximum depth of 18 m (60 ft) (Woo, 1979b) (figure 6.1). The swamp is the result of an impermeable marl layer, approximately 0.5 m (2 ft) thick, which is generally topped by

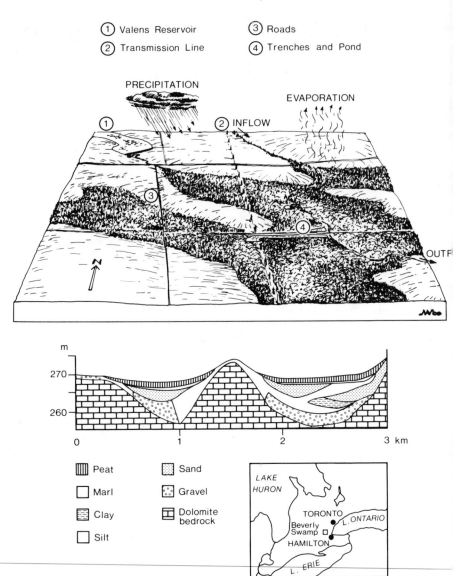

Figure 6.1 Block diagram of Beverly Swamp showing its water balance components and several types of disturbances. Also shown are a geological profile along the Ontario Hydro transmission line and a location map of the swamp.

1 m (3 ft) of peat. Most hydrological activities occur in the peat layer because the groundwater in the clastic sediments does not interact directly with the water in the swamp. The entire area is covered with luxuriant vegetation, and the dominant tree species include white cedar (*Thuja occidentalis*), red maple (*Acer rubrum*), birch (*Betula papyrifera*), and aspen (*Populus tremuloides*).

Beverly Swamp has been considered as a natural water-storage area for Spencer Creek, with ample capacity for the spreading and retardation of water flow (Conservation Branch, 1960). Its natural environment is considered to be worthy of being preserved, and public concern has been expressed regarding developmental plans that infringe upon the swamp (Solandt, 1974). Since the existence and maintenance of a wetland is controlled by its hydrology, a study of the hydrological processes and the impacts of its disturbances will not only benefit the management of Beverly Swamp, but can be put to practical use for other wetlands of the temperate latitudes (Fenton and Welch, 1972).

NATURAL HYDROLOGICAL SETTING

The Water Balance

The annual water balance is determined by the magnitudes of the various sources of water which are added to or taken away from the swamp (figure 6.1). Water is gained from rainfall and snowfall, and inflows in the forms of groundwater and surface runoff. Water is lost through evaporation and outflow from the swamp. In most years, the gains do not exactly balance the losses, and there will be changes in the water storage, as is reflected in a net rise or fall of the water table and the soil moisture content.

The precipitation for Beverly Swamp averages 840 mm (33 in) annually, and is evenly distributed throughout the year. Most of the winter precipitation comes as snowfall which melts mainly in March. The mean annual outflow is 460 mm (18 in), but inflow and evaporation data are available only sporadically (Munro, 1979; Valverde, 1978). Based on the data from 1977, a summer water balance was determined (Woo and Valverde, 1981). It was found that most of the annual evaporation loss occurred between May and early September and that evaporation roughly balanced summer rainfall. Evaporation was about twice the runoff in magnitude and provides an important mechanism through which water is depleted in the swamp.

Water Storage and Water Level

The impermeable marl layer isolates the surface peat layer from the regional groundwater flow of the Upper Spencer Creek basin. The swamp exchanges its water with the atmosphere and the upstream and downstream areas mainly through the peat deposits. Thus, the storage capacity of this peat layer, together with the surface depressions in the swamp, determines the maximum amount of water that can be held in storage.

Water-storage capacity of the peat is limited by its porosity – 80-85 per cent for Beverly Swamp. When the peat is quite dry, there will be plenty of available storage space. Even under the dry summer conditions, however, there is always water retained in the pores so that the peat is never dried out completely. In normal situations, there is a saturated zone in the peat layer where all the pores are occupied by water, and a nonsaturated zone above the water table where the soil moisture varies between 50 per cent at the surface and 80 per cent immediately above the water table. Under wet conditions, most of the pores are occupied by water and the water table is then very close to the surface. Any additional water supply will be accommodated by the numerous surface depressions. When these depressions become full, water moves laterally across the swamp as surface runoff.

Generally speaking, the higher the water table, the smaller the remaining capacity for the swamp to absorb additional moisture. The seasonal fluctuation of the water table at a site in Beverly Swamp is shown in figure 6.2. In spring, meltwater releases a large supply of water to saturate the swamp. The water table rises rapidly above the ground and additional meltwater generates surface flow across the flat terrain. As summer arrives, evaporation and lateral drainage reduce the water held in storage, and the water table also drops (Munro, 1984), although occasional rainstorms can cause sharp rises within several hours. In the autumn, decreased evaporation coupled with water released from upstream usually leads to a flooding of the swamp. Flooding continues into the freezing period when an ice layer generally covers all the depressions. Snowfall soon replaces rain as the major form of precipitation but snow does not yield water to the swamp immediately. Reduced water supply causes the water level to fall, leaving an air space between the ground and the base of the ice layer (Smith, 1984). Since both this air layer and the snow cover are good insulators against the winter coldness, the peat layer seldom freezes, and water can circulate freely within the peat (Woo and Valverde, 1982).

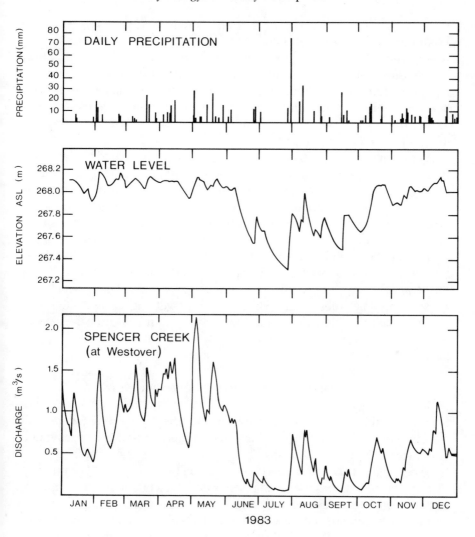

Figure 6.2 Daily precipitation, groundwater level fluctuation at a site in Beverly Swamp, and discharge of Spencer Creek at Westover, 3 km downstream of the swamp

The Swamp as a Streamflow Regulator

Wetlands have traditionally been considered to act like sponges, absorbing water when the supply is large and discharging water gradually afterwards, thus regulating the outflow regime. To be effective as a streamflow regulator, the swamp should have abundant storage capacity to absorb heavy rainfall or snowmelt inputs so that peak runoff is dampened; and during the dry period, water should be released slowly from storage to maintain a moderate level of flow.

A comparison of the seasonal storage change (as indicated by the water table) with the streamflow pattern shows that they follow each other closely. In spring and autumn after the swamp is saturated by the first severe snowmelt or autumn rain events, any additional water will run off quickly. During this period, the regulatory capacity of the swamp is negligible. In the dry summer, the water table is usually low and some of the rainfall can be retained in the peat to be released in the next several days, both to evaporation and to streamflow. It is only during this season that the swamp acts as an effective regulator of streamflow. These results are substantiated by other studies which show that wetland storage is effective when its water table is low, and that wetlands provide little regulatory capacity for streamflow except on a short-term basis (Bay, 1967; Verry and Boelter, 1975).

HYDROLOGICAL IMPACTS OF DISTURBANCES

Several notable man-made changes have occurred in Beverly Swamp, all of which were localized. In the early part of this century, peat excavation left several elongated depressions at the centre of the swamp and these have become permanent ponds (Conservation Branch, 1960). The building of a gravel road across the swamp separated the wetland into two halves, and the road-bed prevents easy water movement from the northern segment to the swamp outlet, except through culverts and along Spencer Creek. In late 1974, the Hamilton Region Conservation Authority dredged a deep hole near the road to create an artificial lake to be used for recreational fishing. This operation required the pumping of sediment-laden water from the pit to Spencer Creek. The suspended sediment concentration of the creek rose from a natural level of 10 parts per million (ppm) (by weight) to a peak of 103 ppm observed on 11 November 1974. When excavation ceased, the pit was rapidly filled by groundwater discharge, and the sediment content of Spencer Creek fell back to its normal level. The swamp was able to recover from this local disturbance quite quickly.

A more major change was related to the erection of fifteen towers to support a 500-kilovolt transmission line across the swamp. Each tower has four legs anchored to the dolomite bedrock. A comprehensive impact study was undertaken by Ontario Hydro to monitor the nature and the extent of environmental changes along the transmission corridor through Beverly Swamp (Fenton and Welch, 1982), and the following is a summary of the results (Woo, 1979a, 1979b).

Three phases of changes can be recognized, including the site-preparation stage (1975), the construction years (1976-7), and the recovery period immediately after tower erection (1977-8). Site preparation requires the clearing of the forest to create openings for the future towers and to establish corduroy roads (large logs placed upon the access route) for vehicular movement (figure 6.3). The loss of trees reduced the interception of precipitation so that there were local increases in rainfall and snowfall reaching the ground. The openings also received more radiation and the resulting differential heating between the forest and the clearings produced

Figure 6.3 An opening in the swamp cleared for the construction of a Hydro transmission tower. The corduroy road around the tower supports the weight of vehicles used during the construction phase. Photograph was taken in March 1978 when meltwater saturated the peat and filled up the surface depressions.

more rapid springmelt and higher summer groundwater temperatures at the open sites.

Under natural conditions, the impervious marl layer separates the highly alkaline groundwater in the clastic materials from the more freely circulating groundwater in the peat. Construction at the tower sites can upset this situation as several types of disturbances are introduced, including test drilling, piling, pumping, the insertion of pipes, grouting, setting of steel plates, and tower erection. These activities punctured the marl, disturbed the peat structure, and altered the surface topography. The result was a rise of the groundwater from the confined aquifer, causing a rise in pH and increasing the water hardness around the tower sites. To minimize contamination of the perched groundwater, Ontario Hydro pumped the highly alkaline water into large tanks and discharged it back through pipes into the deeper aquifer.

The operations of drilling, grouting, and pumping produced uneven water-table elevations at the construction sites. In areal extent, the disturbed sites occupied only a small portion of the swamp. The localized construction effects were easily diluted when autumn and spring floods swept across the disturbed zones. The corduroy road was quite successful in reducing the impact of the vehicular traffic which moved heavy equipment through the swamp. This road did not impede or concentrate water movement and was able to preserve the general flow pattern in the swamp.

Owing to the limited extent of the disturbances and the effort of Ontario Hydro to reduce the impacts, the swamp could absorb the changes without serious hydrological consequences. The natural swamp environment also enhanced the recovery. One factor is the ability of the peat to store and to transmit water so that short-term fluctuations in water level induced by pumping or by excess rainfall and differential evaporation between the open and the forested sites can be evened out quickly. Another factor is the rapid regeneration of vegetation at the disturbed sites (figure 6.4). Finally, the flood events allowed the surface runoff to remove any hydrological effects, in terms of both the quantity and the quality of water, from the local sites. Since flooding is recurrent in a swamp, the construction effects were greatly reduced within several years.

CONCLUSION

In its natural state, a temperate-latitude wetland undergoes recurrent flooding, usually in both spring and autumn. Although the peat layer has a large water-holding capacity, it cannot regulate storage and runoff ef-

Figure 6.4 The same site photographed in September 1985 showing the luxuriant regrowth of vegetation

fectively if the water table in the peat is high. Thus, such wetlands are seldom effective in providing long-term storage. In the dry season, however, evaporation is considerable and the water table often remains quite low. The wetland will be able to absorb heavy rainfall and gradually release this water to runoff over a period of days.

There remains, at present, only a small fraction of the original coverage of wetlands in southwestern Ontario. Beverly Swamp is an example of a wetland that has been encroached upon by rural development over the past two centuries. Recent disturbances to this swamp have left their imprints in the form of roads, ponds, and a power-line transmission corridor. These disturbances were localized in extent and did not seriously affect the wetland hydrology. Other Southern Ontario wetlands, however, have been drained and reclaimed for agricultural and urban land usage. Snell (1982) estimated that about 80 per cent of the Southern Ontario wetlands have been thus destroyed. Their water table no longer rises frequently to the ground surface and the original peat structure is permanently altered. A change in the hydrological environment will irreversibly alter the wetland ecology. Should wetlands be considered as a desirable refuge for the wildlife in Southern Ontario, their continued presence will require that the natural hydrological environment be preserved.

REFERENCES

Bay, R.R. 1967. Factors influencing soil-moisture relationships in undrained forested bogs. In Sopper and Lull (1967): 335-42

Conservation Branch. 1960. *Spencer Creek Conservation Report.* Department of Commerce and Development, Toronto

Fenton, W.W., and Welch, W.E. 1982. Transmission corridors and wetlands – a review. *Proceedings of a Pre-conference Session of the Ontario Wetlands Conference*, September 1981, Toronto, 163-78

Munro, D.S. 1979. Daytime energy exchange and evaporation form a wooded swamp. *Water Resources Research* 15: 1259-65

– 1984. Summer soil moisture content and the water table in a forested wetland peat. *Canadian Journal of Forest Research* 14: 331-5

Smith, S.L. 1984. Ground and water temperatures in a mid-latitude swamp. MSc thesis, McMaster University

Snell, E. 1982. An approach to mapping the wetlands of southern Ontario. *Proceedings of a Pre-conference Session of the Ontario Wetlands Conference*, September 1981, Toronto, 1-26

Sopper, W.E., and Lull, H.W., eds. 1967. *Forest Hydrology.* Pergamon Press, Oxford

Solandt, O. 1974. *Report of the Solandt Commission: A Public Enquiry into the Transmission of Power between Nanticoke and Pickering.* Queen's Printer, Toronto

Valverde, J. 1978. Water level regimes in a swamp. MSc thesis, McMaster University

Verry, E.S., and Boelter, D.H. 1975. The influence of bogs on the distribution of streamflow from small bog-upland catchments. *Hydrology of Marsh-Ridden Area.* Proceedings of the Minsk Symposium, UNESCO Publication no. 105: 469-78

Woo, M.K. 1979a. Impact of powerline construction upon the hydrology of part of a mid-latitude swamp. *Catena* 6: 23-42

– 1979b. Effects of power line construction upon the carbonate water chemistry of part of a mid-latitude swamp. *Catena* 6: 219-23

Woo, M.K. and Valverde, J. 1982. Ground and water temperatures of a forested mid-latitude swamp. *Proceedings, Canadian Hydrology Symposium: 82,* June 1982, Fredericton, 301-12

– 1981. Summer streamflow and water level in a mid-latitude forested swamp. *Forest Science* 27: 177-89

PART TWO

THE BUILT
ENVIRONMENT

The old Farmers' Market
(from W.H. Carre, *Artwork on Hamilton, Canada*, published in 12 parts, 1899)

Darnley Mill
(photo by R. Bignell)

The previous city hall
(from W.H. Carre, *Artwork on Hamilton*, *Canada*, published in 12 parts, 1899)

Ethnic Hamilton: Little Portugal
(photo by R. Bignell)

Residential area, circa 1920
(photo by R. Bignell)

Westdale: an early suburb
(photo by R. Bignell)

CHAPTER 7

The beginnings: Hamilton in the nineteenth century

R. LOUIS GENTILCORE

THE TOWNSITE (1788-1825)

From 15 January to 12 March 1788, a number of survey lines were drawn, setting in motion the series of events that would result in the establishment of Hamilton. The lines defined Lot 14 in concessions II and III of Township no. 8 (later Barton Township) at the head of Lake Ontario. Three decades later, the same lot appeared as the townsite on an early map of Hamilton (figure 7.1). The northern portion is divided into twelve 0.8 ha (two-acre) blocks, ten of which are subdivided into town lots around two reserved for public use. The townsite at Hamilton was neither unique nor imaginative. There was little to distinguish it from dozens of others around it, each aspiring to leap into urban maturity. But this ordinary plot of ground had undergone significant change. Already it contained the seeds of growth that would bring forth a major urban place – the second one in the province and the fourth in the country before the end of the century.

A careful reading of the map reveals some of the 'seeds of growth' – names of features, places, and individuals which would play a major role in the transformation of the townsite. The townsite itself is bounded on the north by 'Present Road from Queenston to Burlington' (King Street), on the east by 'Road to Hughson Landing' (James Street), and on the west and south by King and Hunter streets. The third concession road (Concession III is south of King, Concession II is north) is Main Street. The 'Road up the Mountain to Ancaster' later became John Street. This simple description identifies a major feature of the city's development: the continuity of urban form. The townsite in the wilderness is recognizable as the continuing focus of urban activity to the present day.

Another set of names identifies the prime movers in the struggling village's early development. 'Bellevue' was the first substantial house on

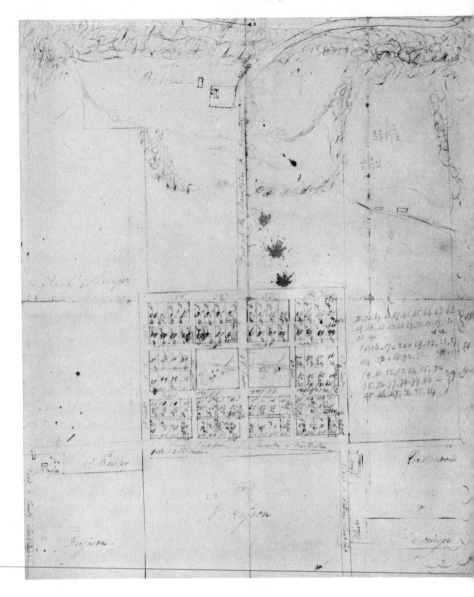

Figure 7.1
[1820?] Early plan of Hamilton. Original in Metropolitan Toronto Library

Lot 14, a two-storey stone mansion with floor-length windows and elegant verandahs. The house was built in 1805 by James Durand, a recent arrival from London who had purchased Lot 14 (in Concession II) as a base for business operations in the province. Durand also laid out the first townsite, between King and Main streets, on land purchased from another prominent landowner named on the map, Nathaniel Hughson. Hughson had come as a loyalist, receiving as a grant that part of Lot 14 extending from Main Street to the bay. Together, Durand and Hughson promoted the townsite as the location of a projected new court house. The agitation was continued successfully by George Hamilton, following his purchase of Durand's property in 1815. (Short biographies of Hamilton, Hughson, Durand, and other prominent early Hamiltonians will be found in Bailey, 1981.)

Also shown on the map are the two blocks of land offered to the Crown for public buildings. The key building, the Court House, was in place by 1817. The log-and-frame building, with the court room above and the jail below, lasted ten years. Each of its replacements has continued the court function on the same block to the present day.

The Court of Quarter Sessions of the Peace was part of the legal apparatus set up to administer the new lands of western Quebec (later Upper Canada). As population increased, the 'upper country' was divided into districts for judicial purposes. For each district, a judge, justices of the peace, a sheriff, a court clerk, and a coroner were appointed. The justices of the peace, or magistrates, bore the brunt of local administration. As government appointees 'on the spot,' they administered local finances, road building, the construction and operation of jails and court houses, the regulation of ferries, the sale of liquor licences, and the appointment of local officials. Possession of the court function conferred tremendous advantages for growth on any aspiring young settlement.

In 1816, the District of Gore was created at the head of the lake, from parts of the existing older districts of Home and Niagara. The naming of the district had been anticipated for a number of years but the selection of the district centre was another matter. Competition mounted for the prize. The credentials of Hamilton (Durand's farm) were not very impressive, in contrast to those of its rivals – Dundas (Cootes Paradise), Crook's Hollow (above Dundas), Brant's Block (Burlington), and Ancaster.

Dundas had begun on the townsite planted at the junction of Spencer Creek, at the very head of Lake Ontario, and the newly surveyed Governor's Road (Dundas Street). A second settlement followed nearby at a good mill site on Spencer Creek. The combination spawned a thriving milling and shipping centre, with a population in 1812 of more than two hundred. Another industrial area, just above Dundas on the escarpment brow, had political pretensions of its own. This was Crook's Hollow, utilizing the power of the upper Spencer Creek to drive grist and saw mills to which were later added a brewery, a distillery, oil mills, a cooperage, and a woollen mill.

A short distance away, on the north shore of Lake Ontario, a large tract of land had been awarded to Joseph Brant, the Mohawk loyalist. In 1810, James Gage, a merchant and farmer in Stoney Creek, purchased a portion of Brant's Block. He laid out a townsite, Wellington Square, to which he added a steam flour mill, with wharf and warehouse. A leading grain market at the time of the War of 1812, Brant's Block also petitioned for the court house.

In terms of size and prestige, however, the leading contender was the thriving village of Ancaster. With a population close to 2,000 in 1817, Ancaster had been the largest milling centre in the area and was clearly the commercial and industrial centre of the area formed as the Gore District. Its major disadvantage for administration was its awkward position above the escarpment, removed from the lakeshore.

In spite of the claims of its competition, Durand's Farm was chosen as the site of the court house and the new district centre and christened after its new owner, George Hamilton. Ancaster and Dundas countered with renewed efforts to have the decision changed, but the more effective lobbying ability of Durand, Hughson, and Hamilton had won out. Equipped with its new function, Hamilton was ready to establish itself as the leading urban place at the head of the lake.

In some ways the geographical attributes of the new district centre were not promising. It was not on Simcoe's road from York through Dundas to London, the projected principal axis of the province. It was shut off from the lake by the Burlington bar (the Beach Strip). Most of the lakeshore was low and swampy. Inland, expansion would be hindered by steep slopes – terraces cut by numerous ravines and the formidable bulk of the escarpment.

At the same time, a number of regional advantages were available and could be made to work in Hamilton's favour. As the formation of the new district recognized, the appearance of farms and mills had marked the head of the lake as a growing area with potential for development. Critical to this development would be the improvement of transportation facilities. The head of the lake lay at the junction of two major continental corridors – the Great Lakes–St Lawrence route and the Mohawk Valley–Niagara

Peninsula route. As early as the 1780s, the vanguard of settlement had moved into this area. In 1790, Richard Beasley set up a store and trading post on the strategic Burlington Heights, the high gravel bar at the western end of Hamilton Harbour. The War of 1812 confirmed the pivotal role of the bar. Most of the fighting in the area was concentrated in the Niagara Peninsula, the neck of the land between lakes Ontario and Erie. The major campaign movements – the westward retreat of the British forces from Niagara to Hamilton, the American attempt to cut their supply lines at Stoney Creek, and the readvance from Burlington Heights to Niagara – stressed the importance of the east-west route that ended at Hamilton. Indeed, military concerns may have been the most telling argument in Hamilton's petition for the district centre.

A return to this map (figure 7.1) enables us to see the emphasis put upon routes at this time. First, the route discussed above is identified (the 'Present Road from Queenston to Burlington') and shown relative to the townsite. Second, another route, the 'Road Up the Mountain to Ancaster,' indicates an appreciation of the future hinterland; it may also suggest that the escarpment was not viewed as the major obstacle that we now assume it to have been.

THE TOWN (1825-42)

Hamilton's hold on the court function was more substantially fixed in the 1820s. In 1827, construction began on a new stone building for the court. In the same year, Hamilton became a port, enabling it to compete as a marketing centre with its rivals, Dundas and Ancaster. As district town and port, the small village was able to draw upon and serve hinterlands to the west, south, and north, initiating undertakings that would mark the city's growth to the end of the century.

Hamilton became a port with the excavation of a permanent channel through the Burlington bar. The canal era had been ushered in by the construction of the Welland Canal which removed a major barrier (Niagara Falls) on the Great Lakes–St Lawrence route. The channel through the bar opened the head of Lake Ontario to the route, cancelling out the first impulse to reach the open waters of the lake from Dundas. The bustling mill town had formulated a two-part project – a canal (later, the Desjardins Canal) through the marsh and the channel through the bar – that would transform it into a lake port. Unfortunately for Dundas, the latter step was taken first. The bar was cut, giving Hamilton access to open water, while Dundas was still building its canal.

The cut or 'canal' through the bar was nothing more than a dredged channel replacing the natural ditch that was here – a shallow, unreliable passage that forced transshipment of goods to the bayshore by small schooners and barges. Under these conditions, other port sites, such as Brant's Block (Burlington), had been beyond the bar. The new channel transformed the region's spatial relationships, initiating a new set of rivalries. Hamilton was now able to exploit its advantages of direct access to the lake on one side and to a hinterland above the Niagara escarpment on the other. The district town became an entrepôt, receiving and dispatching cargo, sorting, selling goods, and promoting trade. Mill sites nearby declined in importance; accessibility had become the dominant factor in development.

The effect of the new channel can be seen on the map of 1842 (figure 7.2). Four wharves have been built on the harbour (Burlington Bay), three of them on the snout of the well-drained, gently sloping Iroquois sand spit. The major change was at the foot of James Street which had become the town's north-south axis. The fourth wharf, 'Land's Wharf,' emphasizes the attraction to shipping of one of Hamilton's leading land-and-commerce entrepreneurs – the Land family (see Bailey, 1981). Beginning with grants made to them as loyalists, the family had accumulated land from the bay to the escarpment, between the present Wellington Street and Sherman Avenue. One of the Lands (Abel) had organized a shipping business, handling salt, whiskey, potash, grain, and flour from his wharf at the foot of Wellington Street.

The map of 1842 (figure 7.2) shows the young community functioning with two nuclei – the older townsite and the younger port. The latter had its own community, reflecting its role as a reception area, with boarding houses, a hotel, and other facilities for a port population – all of this separated by distance (approximately three kilometres or two miles) and topography (ditches and ravines) from the centre around the court house. However, the map also indicates that the two parts were merging with the stronger pull, significantly, coming from the port side. The point is illustrated by the appearance of the Anglican church (Christ Church) on this side of the town. Begun in 1835, the church was an attractive wooden building, with a classical portico, tower, and spire (Campbell, 1966). Its proximity to police board president Judge Thomas Taylor's prominent house and gardens suggests one factor in the site selection. More significant, however, was the role of a number of Hamilton's élite families. Richard Beasley, Allan MacNab, Robert Land, George Hamilton, and Nathaniel Hughson were charter members of the church. The subsequent

Figure 7.2 1842. Plan of the Town of Hamilton District of Gore, Canada 1842.
Hamilton Public Library

purchase of church pews produced a church layout that can be read as a
map of social status in Hamilton in the 1830s and 1840s (Doucet and
Weaver, 1984).

Hamilton and Hughson had competed for the honour of providing the
site. The decision to accept Hughson's donation of land not only enhanced
his social prestige but also promoted his economic activities. In 1835, he
began the subdivision of his land-holding north of King Street, between
James and Mary. Several city streets still bear Hughson family names –
James, Rebecca, Catharine, and Hughson. Also, at this time, the Gore
appears. The open area was originally conceived as a large square to be
formed by two pieces of land donated by Hamilton and Hughson, between
their town lots and the diagonal route of King Street. Hughson's unwill-

ingness to match Hamilton's donation resulted in the wedge shape that would become an enduring feature of Hamilton's townscape.

Real estate development was also responsible for other townscape features. After 1821, George Hamilton bought back the public square he had donated in 1816. On the 1842 map (figure 7.2), this two-acre block in a prime location is shown subdivided into lots. Like Hughson's, Mr Hamilton's public spiritedness could easily be overruled by prospects for material gain. Also shown on the map is another Hamilton donation designed to promote development – the 'market' on John Street, south of the Court House. The location, however, was not a popular one. Uphill from the town centre and too far from the port, it survived as a haymarket. The main marketplace was established north of the Court House, at the major road junction of York and James streets. Before the end of the decade, a market house and town hall established a physical presence that would endure on the same site to the mid-twentieth century.

The increase in port activity soon extended to improvements in land transport. The major entry road, York Street (the road from York, the provincial capital), is particularly prominent on the map. So are a number of others, spurred on by the private purchase of road allowances. In 1855, James Street was opened to the mountain brow, joining John Street as a through route from the bay to the escarpment. Beyond, these roads continued to hinterland locations at Ancaster, Brant's Ford, and Caledonia. The presence of an inn on the escarpment brow is instructive.

The advent of the toll road or turnpike, established by private companies, promoted the construction and repair of main routes. (Toll roads did not disappear until the late nineteenth century.) Indications of both turnpikes and toll gates are on the 1842 map (figure 7.2). Leading companies built the Dundas and Guelph Road, via York Street; the Waterloo Road, via King Street; and the Hamilton–Port Dover Plank Road, via James Street. The Hamilton-Toronto Plank Road followed in the late 1840s. At the time of the map, three stagecoach services were operating – to Brantford, Niagara, and Toronto. The construction of roads was facilitated by the wealth of quarries in the area, drained plank or gravel roads soon replacing corduroy roads. Macadamized roads appeared soon after the date of the map. John Street was macadamized in 1844, King Street in 1847.

The Hamilton of 1842 was the result of more than land promotion and transportation improvements. Associated with these developments was a significant immigration. From 1834 to 1841, Hamilton, 'the place where the immigrant took breaks,' increased 2.5 times in population, from 1,367 to 3,414. In 1840, 220 houses and shops were assessed at £60 or more (Weaver, 1982). The immigration was part of a major transatlantic movement. From 1820 to 1834, almost 200,000 immigrants from the British Isles entered British North America via Quebec City. The population of Upper Canada increased from 150,000 (1825) to 450,000 (1841), transforming the province and its regions. The Gore District directly recruited some of its potential inhabitants, sending its own agent to Britain in 1840 (Weaver, 1982).

The migrations fed a booming economy. Imports for the new settlements, combined with the handling of increased exports of timber, wheat, and flour, led to a burst of merchant activity in the bustling district centre. The building of the Burlington bar canal attracted a sizable work force. Soon, the demand for settlers' effects spawned manufacturing activity. A major item that emerged and grew into a significant industry was the making of stoves, marking the beginning of Hamilton's association with iron and later steel manufacturing. By 1842, stoves, pumps, and agricultural implements were being produced for a burgeoning market. The first threshing machine in Upper Canada was made here, in a foundry at the corner of James and Merrick streets. Most of the plants were in the town centre, interspersed with other urban functions, or spread along King Street. The establishments were small, often in the manufacturer's home or on his residential property. The variety of production was impressive – guns, augers, sheet-iron ware, stoves of various kinds, brass and tin ware, wagon boxes, carriages, wagons, and sleighs (Campbell, 1966).

Economic vitality was followed by political development. The Act of Incorporation of 1833 established boundaries that can be identified on the map of 1842 (figure 7.2) – the Bay, the escarpment, Queen Street on the west, Wellington (not mapped) on the east. Bush appears east of Wellington and farmland east of Mary Street and north of Barton Street. The new city charter in 1846 would extend the eastern boundary to Emerald Street and the western to Paradise Road.

At the same time, the developed areas within the boundaries exhibited some well-marked area differentiation. Parts of the town were perceived as distinct by Hamiltonians themselves. Doucet and Weaver (1984) recognize certain 'levels of development' at the time, based on commercial and residential functions. The town core was well established. Incorporating the Court House Square, the main market on York Street, and the haymarket on John Street, the highly visible commercial centre would persist as a feature of urban development in the mid- and late-nineteenth century (Davey and Doucet, 1975).

Beyond the core, two distinctive areas had emerged, the North End and Corktown. The former, separated by distance and topography from the town centre, took on a character associated with its wharfs and waterfront. The area survived as a distinct entity to recent times. So did Corktown, which is the only neighbourhood named on the map. The area was subdivided into lots for low-income houses – small frame buildings and shanties – which drew a population dominated by Irish Catholics. The location was an undesirable one in a low-lying area subject to flood. In contrast, high society, or estate housing, had developed in other areas. The earliest of these was the high land below the escarpment where George Hamilton and Peter Hamilton had their houses. A second favourable area was the high land of the Iroquois bar, parallel to Burlington Bay. Two properties here, prominently displayed on the map, deserve special comment as specific expressions of the achievements of Hamilton's merchant princes.

The first of these was Sir Allan MacNab (see Bailey, 1981), whose imposing Regency mansion, 'Dundurn' still overlooks the bay and the town from its commanding site on Burlington Heights. Hamilton's leading citizen had come to the fledgling village in 1822 as its first lawyer. His main law office was at James and King streets with another to follow on the waterfront at MacNab and Brock streets. He was an active participant in the town's boom of the 1830s, particularly active in transportation projects such as roads, the canal, and later railways. In 1837, he owned twenty town lots, eight houses, a wharf, a warehouse, and a tavern.

Adjacent to Dundurn, also on York Street, was the estate and mansion ('Fairlawn') of MacNab's most aggressive rival, Colin Ferrie (see Bailey, 1981). The Ferrie family had established branches of their Glasgow business in Montreal in 1824, with extensions to the head of the lake area. A store in Hamilton on King Street was followed in the 1830s by a series in the Hamilton hinterland – in Brantford, Galt, Nelson, Dundas, and Waterloo. Engaged primarily in the import trade and the handling of wheat and flour, Ferrie also founded a steamship company, with service to Rochester, New York; he helped found Hamilton's first bank (the Gore Bank in 1835) and was its president from 1839 to 1854; he was later to be the city's first mayor, in 1847.

A REGIONAL CENTRE (1851-61)

In 1851, Hamilton's population had reached 10,000. Six years later, it had exploded to 25,000. A major economic boom had transformed the frontier town into a regional urban centre with metropolitan pretensions. The key was transportation. Hamilton saw itself as the hub not only of its region at the head of the lake but also of a continental system (figure 7.3). The major ingredient promoting change and the dream of dominance was the railway.

Railway planning had begun two decades before. In 1834, a group headed by Allan MacNab had obtained a charter to build a railway from Hamilton to London, a proposal that did not survive the uncertainties of the time. The scheme was part of a more general concern with better transportation, including canal construction and road improvmements, spurred on by growing demands for bulk products such as wheat and timber. Between 1851 and 1857 wheat production in Upper Canada approximately doubled. Hamilton, at the head of navigation on Lake Ontario and with a prosperous hinterland, saw the railway as a means of strength-

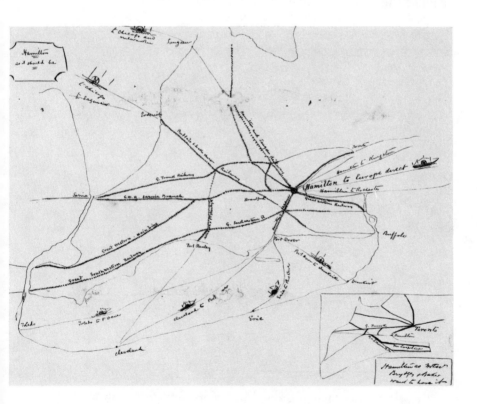

Figure 7.3 1860. Hamilton as It Should Be. National Map Collection.

ening and extending its hold on the marketing function. In the early 1850s, MacNab was the president and director of five projected railway companies (Campbell, 1966).

The main achievement of railway mania in this area was the building of the Great Western Railway. Promoted by Hamilton businessmen who raised their capital in London and the United States, the Great Western was very much 'Hamilton's railway.' It was built through Hamilton, by-passing rival Dundas and Ancaster; making use of the upper bar to connect the north and south shores of the lake, providing the best transport westward (to London); and solidly establishing the town as the transport focus of its region. The changes wrought are illustrated on the Keefer map (figure 7.4). The pre-railway competitive position of Dundas is obvious. Highways provided access east and south of the Niagara Peninsula; west to Brantford and Lake Ontario; and north by the Guelph Road. The railway changed all this. The regional routes remained, but Hamilton became

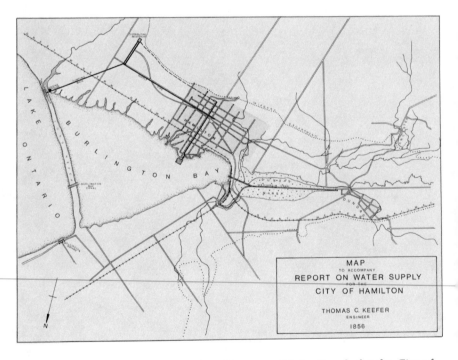

Figure 7.4 1856. Map to Accompany Report on Water Supply for the City of Hamilton. Thos. C. Keefer, Engineer. National Map Collection

their focus, dominating the upper roads from the Dundas Valley and Burlington Heights and the lower road from Burlington Beach. Collecting wheat and other produce from the Niagara Peninsula and southwestern Ontario, Hamilton would also vie with Toronto in attracting trade from the upper lakes by rail from Georgian Bay ports. It also became a major centre on the U.S. immigration route from the east coast (New York and Boston) to the midwest (Chicago and Milwaukee).

The railway brought new activity to the young port, the Great Western grain elevators emerging as the most imposing symbol of Hamilton's place in the grain trade. Cartage agents and freight depots appeared in the town centre. The largest of these was Hendrie and Company at King and MacNab streets. At the height of its operations the firm occupied a whole block, from Bay to Caroline streets, given over to smithies, wagon shops, company houses, stables, and a work force that included up to 300 horses (Campbell, 1966).

The coming of the railway not only reorganized transport but also introduced the age of steam power. Water power had been the basis of manufacturing in the mill towns of Dundas and Ancaster. Steam engines, using coal brought in by ship or by rail, became the new source of power for foundries and machine shops and the making of steam engines and agricultural machinery. But the major industry was the railway itself. The Great Western shops, built in 1849 to manufacture rolling stock, were the first in Canada, making it possible to replace or finish imports from England and the United States. Locomotives were unloaded from schooners and assembled by English mechanics at Hamilton, Stoney Creek, Winona, and other points along the line. In 1863, the Great Western built its first rolling mill – the first in Ontario – in Hamilton, to reroll English rails to withstand Canadian winters.

The boom of the early 1850s was brought to an end by the resounding crash of 1857. Caught in the throes of an international depression and bedevilled by its own excessive railway involvement, Hamilton's population decreased by 20 per cent from 1858 to 1862 (Weaver, 1982). The city's industries reeled under the blow. Most endured and some later expanded. In 1861, Hamilton, with a population of 19,000, had 2,225 employees in a manufacturing economy which looked to the past and to the future. The former included establishments of the preceding decades – textile plants, tanneries and leather-goods firms, brickyards, carriage and wagon works, breweries, distilleries, food-processing firms, lumber and planing mills, flour mills, stove manufactories, and small foundries. The second, and much smaller group, looking ahead to new developments, were the heralds of an association with the railway and steam power. The products were locomotives, railway equipment, steam engines, boilers, and agricultural machinery (Roberts, 1964).

In terms of distribution, the first group was concentrated in the townsite; 85 per cent of the firms were within one-half mile of King and James streets, in the commercial core of the city. This was the 'central manufacturing district' (Roberts, 1964), with small establishments catering to the local population, including merchants from nearby centres. The second group had a different orientation. The assembly of bulk raw materials, such as coal and pig-iron, and the shipment of products beyond the region favoured locations away from the centre. Space and the improved access offered by the railway became important considerations. The experience of the Hamilton Agricultural Works illustrates the change. This firm, which built the first threshing machine mentioned above, was first located at the corner of James and Merrick streets. In 1854, it was rebuilt on Wellington Street, north of the railway on the eastern edge of the city. Also indicative of developments to come was the appearance of oil refineries, making illuminating oil, naphtha, and machine oils from crude oil brought in from southwestern Ontario. These were located on the railway, east of Wentworth Street, outside the city.

Another indication of Hamilton's growth as a regional centre was its ability to assume responsibility for public services. A police force, sewers, street lighting, and water supply not only met the needs of Hamiltonians; they also projected an urban image to the outside world. The construction of a magnificent waterworks plant symbolized the city's ability in both spheres. The need for water for future growth initiated both a search throughout the surrounding area and intensive political lobbying by the city. The result was the passage in 1856 by the province of the Waterworks Act, a piece of legislation with far-reaching geographical consequences. The act empowered a municipality to go outside its boundaries to provide an essential service for future development. In its search for water, Hamilton was allowed 'to enter upon lands within twenty miles [of its limits] and to appropriate as much as may be required' (Johnston, 1958). Hamilton was allowed to construct its own waterworks and to draw water from nearby Ancaster, one of a number of possible sources. Dissatisfaction with Ancaster water led to the tapping of Lake Ontario, an undertaking that required unheard of expenditures – £300,000 for pumps and pumping stations. By 1860, construction was completed on the first project of its kind in the province. The Keefer map (figure 7.4) shows the main components. The first unit was a 360-metre 1,200-foot basin, 5 m (16 ft) deep, excavated on the lake side of the beach strip; this was part of the facilities bringing in water from the lake, filtering it through the sand of the bar. Next, the water was moved by pipeline to the pumping station; it was lifted to Ottawa Street, then south to the Barton Street storage reservoir, 54 m (180 ft) above Lake Ontario. From here the main distribution line proceeded along Main Street to James Street, serviced by auxiliary mains and one hundred hydrants. One earlier suggestion had been to draw water from the bay to a reservoir on the high level bar, at Dundurn – a much more modest undertaking. Both Thomas Keefer, the consulting engineer, and the city fathers agreed that the bay was an inadequate source of water for a future city of 50,000 – a population that was reached before the end of the century. At the time, the reservoir was far removed from the city limits at Wentworth Street. For the next seventy years, as the city spread to the site, pumphouse and reservoir regularly supplied the needs of the growing population. Today, the former remains intact (James and James, 1978), an engineering marvel that is perhaps the city's most impressive monument to its emergence as a major urban centre.

Other monuments persist. Hamilton's legacy of fine mid-Victorian buildings not only reflects the city's achievements in this period but also provides a physical continuity to the present time. Only a few of the most impressive buildings, all built in the 1850s will be cited (Hamilton-Niagara Branch, 1967). The private residences include: Whitehearn, a Georgian town house on Jackson Street, associated with foundryman Calvin McQuestern; Inglewood, a Gothic Revival mansion, built on the heights below the escarpment on the residential estate begun by George Hamilton; Sandyford Place, an Italianate stone terrace block, an example of the first row housing being built by Scottish masons south of Main Street.

The 3-1/2-storey masonry Commercial Block on MacNab Street is a striking reminder of the many large merchant-owned and -operated warehouse and supply buildings located in the Hamilton core. The Custom House, built opposite the Great Western Railway station, was Hamilton's grandest public building, a full-blown proclamation of the city's trading prominence. The Classic Revival Central Public School, on Hunter Street, was the first large graded school in British North America, taking in up to 1,000 students, and was a pioneer in new methods of public-school instruction. Among Hamilton's churches, the most impressive was the Decorated Gothic St Paul's on James Street South, .'a Gothic sermon in stone' (Hamilton-Niagara Branch, 1967). Originally named St Andrews, the church speaks eloquently of the place in Hamilton society of the Scottish Presbyterian élite.

THE INDUSTRIAL CITY (1871-91)

In 1891, Hamilton's population reached 50,000, doubling since 1871. It was the fourth largest city in Canada, a rank it would not retain or regain in the next century. A spatial spread matched the increase in numbers. Most significant was the eastward movement. The city boundary, formerly at Wentworth Street, was now at Sherman Avenue. Between the two streets the city met the countryside. Rural estates, with fenced mansions set in a parkland of cultivated land and trees, mingled with the advance of new surveys and lines of streets (figure 7.5). Elsewhere, consolidation of older sections had occurred and old boundaries remained. To the west, the Chedoke Ravine had not been breached; Dundurn Street still marked the limit of the built-up area. North and south, the escarpment and the bay compelled a lateral eastward expansion, dramatically portrayed in the 'birdseye view' of 1893 (figure 7.5).

For the most part, the institutional geography of the city remained in place. The Court House, reconstructed, was in its original location on Main Street, the principal east-west axis of the city. The Town Hall was

Figure 7.5 [1893. Hamilton, Ontario], print

still at the junction of the road from York (York Street) and the main north-south thoroughfare, James Street. The business quarter was at the junction of James Street and Hamilton's oldest street, the King Street trail. The Gore Park, with its horses and carriages, provided a unique city landmark.

At the same time, new commercial developments had taken place. Most significant was the massing of activity, maritime and industrial, at the bay front (figure 7.5). The shoreline itself had undergone change – a harbinger of more extensive changes to come. The filling in of the bay began in the 1850s with the siting of the Great Western Railway roads, west of Bay Street. Later, many of the long inlets between Mary and Wellington streets were also infilled. One of these, the former eastern creek (near Wellington), had been used for the Hamilton and North-western Railway line. The line, crossing from Ferguson Avenue eastwards along the bay shore towards the exit from the city along the Beach Strip, connected Hamilton with Lake Erie in the early 1870s. Before the end of the decade, extensions northward to Barrie and Collingwood established Hamilton as a unique handler of timber from the Georgian Bay area.

Increase in population numbers was associated with a growth in manufacturing activity. Facilities established in the 1850s and 1860s – a thriving metals industry, the busy railway shops, imports of coal from Pennsylvania and pig-iron from Britain and the United States – were joined by continuing improvements in transportation, including new railway lines and the enlargement of the Welland Canal in 1877. Spurred on by an aggressive city administration and protection afforded by the federal tariff, Hamilton was particularly successful in attracting capital from the United States. By the late 1880s, the city was dubbed 'the Birmingham of Canada.' A group of English visitors in 1889 wrote the following tribute: 'Of all the places we had visited during our trip to the American Continent, the prettiest, cleanest, healthiest, and best conducted was the City of Hamilton, Canada; and from our inspection of the vast and varied manufacturing industries, its one hundred and seventy factories, with its 14,000 artisans, the large capital invested, and the immense output annually, we concluded it was well named the Birmingham of Canada and has undoubtedly a great and glorious future before it' (Johnston, 1958).

The newer industries of 1861, built around the railway and associated with the advent of steam, increased in number, size, and sophistication in three decades. Production embraced rolled iron, nails, tacks, screws, wire, agricultural machinery, engines, boilers, bridges, tin cans, axles, wheels, and cotton goods.

In 1891, most of Hamilton's manufacturing firms were still located in the central part of the city. Two-thirds of its 150 establishments were traditional industries, processing consumer goods – boots and shoes, clothing, some metals, wood products – for a local market (Roberts, 1964). Even here, however, there were new developments. A case in point was the Tucketts Tobacco Company at Queen and York streets, on the edge of the central area (figure 7.5). Two blocks southward, the proprietor, George Tuckett, could literally look down on his 200 employees, from his mansion Myrtle Cottage, on the high land at King and Queen streets.

The most striking developments were those of other larger firms away from the centre, characterized by a higher degree of specialization and oriented to a wider market. They are identified in figure 7.5 by a heavier density of smoke and smokestacks, in particular on the western bayshore, adjacent to the Great Western railyards. Most prominent here are the Ontario Rolling Mills (no. 20 on figure 7.5), begun in 1879 by Ohio iron interests in the former buildings of the Great Western Mills. Opened just as tariff legislation came into effect and incorporating American technology, the mills, with 500 employees in 1891, were the basis of a new industrial activity that would culminate in the formation of the Steel Company of Canada in the next century. Nearby was the Hamilton Bridge Company (no. 15 on figure 7.5) which had built the swing bridge over the Burlington Canal in 1876. In the same area were firms making nails, including the Ontario Tack Company from Chicago at the foot of Queen Street and the enlarged Greening Iron Works at Queen and Napier streets.

Farther east, along the Great Western Railway, at Wellington Street, another concentration was emerging on land made available by the filling in of a creek bed. The Sawyer Massey agricultural-implements factory, the outgrowth of a foundry dating to the pre-railway period, moved here from a central location in 1857. Tax exemptions offered by the city helped promote the move, as well as that of the Canada Screw Company, founded in Dundas in 1865 but enticed here in 1887. Three of the firms mentioned above – the Ontario Rolling Mills, the Ontario Tack Company, and the Canada Screw Company – were instrumental in the subsequent founding of the Steel Company of Canada.

The textile industry, a major yardstick of industrial achievement, and the earliest fostered by federal tariff policies, was also represented in the new industrial area. The Hamilton Cotton Company (no. 5 on figure 7.5), with 200 employees in 1891, was established here in 1880. North of the Great Western, on James Street, the Ontario Cotton Mills (shown on figure 7.5 but not numbered) had 500 employees. But the largest textile employer remained in the central area. The Sanford Manufacturing Company at the corner of King and John streets, with more than 3,000 employees, was the largest clothing establishment in Canada in 1891.

In summary, manufacturing in Hamilton was still dominated by a large number of small enterprises, oriented to the local market and located in the core of the city. In contrast a much smaller number of firms were on the railway. But there were larger factories, many of them associated with the larger spaces made available by landfill near the bay. At the same time, it should be noted that there were no industries on the bayshore itself. Waterfront sites would not be part of Hamilton's industrial geography until the next century.

The changes wrought by industrial and population growth were reflected in the morphology of the city. Hamilton was no longer a pedestrian city. The east-west distance between Dundurn Street and Sherman Avenue was almost 5 km (3 mi); from the Bay to the escarpment, 6 km (4 mi). Differentiation in the built-up area continued (figure 7.5). Commerce and small industry were concentrated in the core; to the southwest was a cluster of upper-class residences; to the north and east were the railway belts with associated industries. The fragmentation along social and economic lines was particlarly apparent in the city's north end where the city's major concentration of working-class housing mingled with factories, railways, and marshes (Doucet, 1976). At the same time, places of work and residence were being separated, facilitated by the emergence of a street-railway system. Streetcars also enabled Hamiltonians to reach widely dispersed recreational facilities, such as the racetrack on the eastern edge of the city and the Crystal Palace exhibition area at King and Locke streets (Victoria Park on figure 7.5).

Public transport was utilized in 1874, using cars drawn by horses or mules. The first route was east-west along Main Street, along the main axis of the city's lateral spread; by 1891, cars were running well beyond the city's eastern boundary. The second route followed the north-south axis, James Street, serving both the wharf area and the city west of the core. By 1891, when the system was electrified, twenty to twenty-five cars were in service, with approximately 16 km (10 mi) of double track. A different form of 'streetcar' advanced the city's continuing efforts to join the escarpment to the city below. The inclined railway appeared, scaling the escarpment face at a 45-degree angle, reaching the upper terminus in little over one minute. The first incline was concentrated at the head of Jarvis Street (figure 7.5) between 1889 and 1892. The second, at Wentworth Street, although also shown on the map, was not completed until 1899. Both inclines continued in service until the 1930s.

Hamilton's attempts to come to terms with its physical disabilities would continue, hand in hand with renewed industrial growth. By 1891, the legacy of the railway and the industrial age of the 1850s was once again apparent. In terms of North American urban development, Hamilton was still a small town, struggling to adjust its physical dimensions to its economic accomplishments. The most impressive of these was a healthy industrial base, making possible a steady growth in population from 1871 (26,900) to 1891 (48,960) (Weaver, 1982). The growth would continue and was anticipated. The 'Birmingham of Canada' was due to come of age.

REFERENCES

Bailey, T.M., ed. 1981. *Dictionary of Hamilton Biography*, vol. 1. Hamilton

Campbell, M.F. 1966. *A Mountain and a City: The Story of Hamilton*. McClelland and Stewart, Toronto

Davey, I., and Doucet, M.J. 1975. The social geography of a commercial city, ca.1853. In Katz (1975): 319-42

Doucet, M.J. 1976. Working class housing in a small nineteenth century Canadian city: Hamilton, Ontario 1852-1881. In Kealey and Warrian (1976): 83-105

Doucet, M.J., and Weaver, J.C. 1984. Town fathers and urban continuity: the roots of community power and physical form in Hamilton, Upper Canada in the 1830's. *Urban History Review* 13 (October 1984): 75-90

Hamilton-Niagara Branch, 1967. *Victorian Architecture in Hamilton*. The Architectural Conservancy of Ontario, Hamilton

James, W., and James, E.M. 1978. A sufficient quantity of pure and wholesome water: the story of Hamilton's old pumphouse. Faculty of Engineering, McMaster University, Hamilton

Johnston, C.M. 1958. *The Head of the Lake*. Robert Duncan, Hamilton

Katz, M.B., ed. 1975. *The People of Hamilton, Canada West*. Harvard University Press, Cambridge and London

Kealey, G.S., and Warrian, P., eds. 1976. *Essays in Canadian Working Class History*. McClelland and Stewart, Toronto

Roberts, R.D. 1964. The changing patterns in distribution and composition of manufacturing activity in Hamilton between 1861 and 1921. MA thesis, Department of Geography, McMaster University, Hamilton

Weaver, J.C. 1982. *Hamilton – An Illustrated History*. James Lorimer, Toronto

CHAPTER 8

Emergence of the modern city: Hamilton, 1891-1950

HAROLD A. WOOD

THE CITY IN 1891

In 1891, Hamilton was entering a period of new opportunities after a quarter-century of unspectacular if sustained population growth at a nearly constant rate of about 3 per cent per annum. The advantages of its location – a protected harbour at the head of Lake Ontario and a position opposite a major discontinuity in the Niagara escarpment – had failed to bring the expected commercial activity. Whereas the city was dominant within a region including most of the Niagara Peninsula and extending perhaps 48 km (30 mi) to the west, the national and, indeed, international prominence which Hamilton once envisioned had not materialized as this role was pre-empted by Toronto, close by and with vastly superior physical endowments.

Both site and situation had a number of disadvantages for transportation. The break in the escarpment was choked with hummocky glacial deposits, discouraging its use as a corridor of movement, and, as Hamilton was situated on the south side of the harbour, it was bypassed by the main line of the Great Western Railway running westward from Toronto (Wood, 1960). The routes between Buffalo and Detroit were well to the south, on flat land above the escarpment, while the Trans-Canada Railway lines, passing east of Georgian Bay, had their natural regional terminus in Toronto. In terms of railway access, therefore, Hamilton's centrality extended over only a relatively restricted geographical area (Westland, 1951).

The harbour, of course, was useful for shipping, but it was, in fact, too well protected. The bar cutting it off from Lake Ontario has no significant natural break; until shortly before 1891, vessels entered and left the harbour through a narrow canal only 2 m (7 ft) deep.

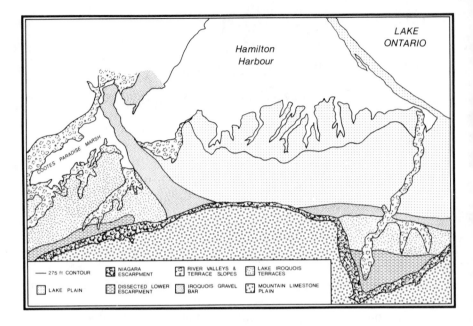

Figure 8.1 Physiographic setting of Hamilton

Evidence that Hamilton's development had been less than dynamic may be seen in the lack of physical expansion. From the date of its incorporation in 1846 until 1891, the city had not needed to take in any new land. It remained comfortably within its original boundaries without having to make those difficult adjustments to site which growth often demands. From the town centre, located then as now at the intersection of King and James streets, the built-up area extended almost equal distances to the east, north, and west. The boundaries of this area corresponded quite well with the city limits except to the northwest where the extremity of the southern branch of the high-level Iroquois Bar was used in part as a cemetery.

To the south, the abrupt 90 m (300 ft) rise of the escarpment constituted an effective barrier to urban expansion but also offered some advantages. The finest residences were built on the lower slopes to escape the congestion and noise of the city. So was St Joseph's Hospital, opened in 1890 (Wingfield, 1946). Higher up, sites could be found for reservoirs at the critical elevations from which water, pumped from Lake Ontario, could flow by gravity into the city at pressures suitable for domestic use.

It was possible to negotiate the escarpment following tracks slanting upwards from the heads of Queen, James, and John streets or in the ease of the James Street inclined railway. Opened in 1890 to facilitate contact with the agricultural area south of the city, the James Street railway was designed more to accommodate wagons than passengers. But the only urban element above the escarpment, or, in local parlance, on the Mountain, was an asylum for inebriates, placed in this isolated location presumably to protect the sensibilities of the more respectable townspeople (*Spectator*, 25 October 1984).

As for land use elsewhere in the city, the most notable aspect was the rather haphazard mixture of industries and working-class houses at lower elevations to the east and north. The harbour, as such, was not used much; industries tended to concentrate instead along the railway lines or near the city centre, but many seem to have been quite indifferent to variations in accessibility. Higher ground, where the Iroquois Bar swings inland from the present shoreline, was essentially residential but still somewhat diversified as it contained both lower- and middle-class homes.

Socio-economic contrasts were extreme. While many families lived in comfort and some in luxury, the working classes experienced great hardship. Most people in this category lived in small rented cottages, poorly constructed and lacking in amenities. Except at King and James streets, where cedar blocks were used for paving, streets were unsurfaced, a condition associated with semi-permanent mud in the lower parts of the city where soils were stone-free silts and clays. Winters were particularly hard, as most factories were unheated and many closed down for weeks or months during periods of severe weather, depriving workers of wages just when their need for cash was greatest. The only public park was the Gore in the town centre, but even this sliver of open space had been available for general use only since 1884; prior to that date it had been fenced in, the gates only opened on special occasions. One can understand why there were so many inebriates that they needed a special facility.

Yet, by 1891 the stage had been set for the transformation of Hamilton. Outside the city, and largely independent of it, changes had taken place which were to have profound effects. The most important was a reconstruction of the Welland Canal, completed in 1887. Locks were increased in depth from 3 to 4 m (10 to 14 ft), in width from 8 to 15 m (26 to 45 ft), and in length from 50 to 78 m (150 to 260 ft) (Petrie, 1967). This improvement greatly facilitated travel by ship from Lake Ontario towards the growing markets of Western Canada and access to such natural resources as the iron ore of Minnesota.

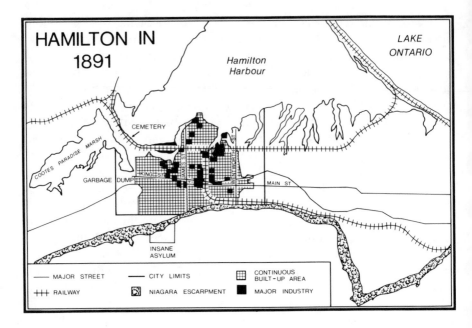

Figure 8.2

A second significant change was the adoption by the Government of Canada in 1890 and 1891 of new legislation protecting national industries from foreign competition. Only after the implementation of these measures and the upgrading of the Welland Canal did it become cheaper to manufacture pig-iron in Canada than to import it from Britain.

For Hamilton, the effects of these changes were, first, that the city become unquestionably industrial rather than commercial, and second, that it was able to attract much investment from u.s. companies, anxious to retain or increase their hold on the Canadian market, but also desirous of keeping their investments as close as possible to the international border.

THE AMBITIOUS CITY, 1891-1914

The decades preceding and following the turn of the century were the most exciting in the history of Hamilton because they were marked by a remarkable outpouring of human energy and innovative initiatives. The impact on population growth was not felt at first; in 1906 the city had only 12 per cent more inhabitants than it had had fifteen years earlier.

But from 1906 to 1911 the population exploded, increasing by more than 53 per cent.

The achievements of the civic leaders included bringing to Hamilton the first telephone exchange in Canada (Walker, 1980) and, through an aggressive promotional campaign, attracting a number of major industries, exploiting with much success the predisposition of u.s. interests to invest in the city. Among several metal-working industries which were established during the early years of this period, one deserving special mention is the Hamilton Blast Furnace Company founded in 1893, and party to a series of subsequent amalgamations leading in 1910 to the formation of Hamilton's largest industry: the Steel Company of Canada (Stelco).

But the most spectacular achievement of Hamilton's entrepreneurs was to develop and bring to the city the cheapest and most reliable supply of electricity available to any large urban centre in Ontario. Previously some small communities near Hamilton had obtained electricity from streams cascading over the escarpment but the streams were small and produced little power, particularly during summer dry spells and in the winter. The Niagara River, of course, had lots of water but was too massive to be easily exploited. Nothing much could be done until the outlook was transformed with the latest improvements to the Welland Canal. To operate the larger locks of the new canal, water was drawn from Lake Erie, flowing by gravity along a new channel to the top of the escarpment. The supply water was virtually unlimited, and the channel capacity was considerably in excess of what was required for the locks. It occurred to a group of Hamilton businessmen to acquire the surplus water, drop it in flumes down the escarpment at a site just west of the canal, and use it to generate power for transmission to Hamilton 54 km (34 mi) away. It was a rather daring project; never before had electricity in such quantity been moved so great a distance. But, fortunately, the operation was a complete success, and by 1898 the arrival of this power in Hamilton made it a city of electricity as well as of steel.

The impact on industry was immediate. In 1896, before the power project was even complete, the Westinghouse Company built its first plant in Hamilton, followed by Otis Elevator in 1900 and International Harvester in 1903, to name just the largest of the new arrivals. Soon another asset was added: natural gas from wells near Lake Erie. By 1913, forty-six u.s. firms had set up branch plants in Hamilton (Walker, 1980). From 1905 to 1915 industrial investment tripled. From 1900 to 1911 industrial employment in Hamilton grew by 107 per cent, twice as fast as in Toronto and eighteen times as rapidly as in the Hamilton of the previous decade (Middleton and Walker, 1980).

The expansion of heavy industry had an effect on the composition of the labour force. For the first time, the city began to attract an appreciable number of workers from continental Europe. Between 1891 and 1911, the percentage of migrants from this source increased from 6 per cent to over 15 per cent (Census of Canada, 1911). Also, because heavy industries employed mostly men, there was an ample supply of female labour, which contributed to the growth of much light industry, such as textiles and food processing. By 1911, one-half of the labour force of Hamilton was employed in the manufacturing sector.

This period in the city's history was also one of unprecedented development in the field of transportation, commencing in 1891 with the introduction of the so-called 'radial' electric cars, running on light tracks. The first line ran to Oakville, but within a few years the service was extended to Dundas, Brantford, and Beamsville (Johnston, 1958). Then, towards the end of the decade, a new railway appeared – the Toronto, Hamilton, and Buffalo (TH&B) – created to provide better connections with the United States, a major function being the transportation to Hamilton of coal from the Appalachians. Entering from the east and descending the mountain a few miles from the city, the TH&B line followed the base of the escarpment, where drainage was good and interference with other land uses minimal, as far as John Street, having sent off a spur line towards the waterfront industrial area across what was then open country in the vicinity of Gage Street. West of John Street, the line had to cross the built-up area of the Iroquois Bar, the land-use conflict being mitigated to some extent by the fact that a tunnel was cut to carry the track through the highest parts of the bar. Emerging into the Chedoke Valley, the main line swung north towards Toronto; a branch line continued to the west, and a freight yard was established at the junction.

At the same time, a more far-reaching innovation appeared: the gasoline-powered motor vehicle. By 1900, Hamilton had its first automobile, and in 1903 the Hamilton Automobile Club was founded, the first such club in Canada. Yet the impact of this technological revolution was slow to develop, largely because of the lack of adequate roads. At the turn of the century, most roads were rutted, often impassable because of mud. Where it existed, road improvement consisted mainly of laying planks across the track, producing a surface which was possibly firm but certainly not conducive to high-speed travel. Not until 1904 did Wentworth County begin to create a network of roads with macadam surfaces; for travel of any distance the train was used, or sometimes, as between Hamilton and Toronto, the steamship. Up to 1914, motor vehicles were curiosities or luxuries more than essential components of the transportation system.

Certainly, Hamilton did not need motor vehicles to grow in area. In 1914 the city's physical size was almost two and a half times what it had been at the beginning of 1891. Most of the expansion was to the east, led by the demand of heavy industry for waterfront sites. In an initial approach to zoning, a strip of land running inland from the shore was annexed specifically for industrial use (Weaver, 1982). The area was divided into a number of separate blocks by a series of long, swampy inlets representing drowned valleys cut by small creeks into the soft deposits of the plain at a time when the lake level had been considerably lower. Although these re-entrants gave the shoreline a rather untidy look, in some respects they were quite useful. East-west rail lines were kept sufficiently far to the south to leave adequate space along the harbour for major industries; the inlets constituted convenient boundaries for individual properties; and, where desired, they could be depended on for the docking of vessels. Nevertheless, they could pose limits to the lateral expansion of sites occupied by the most dynamic industries. In such cases, notably that of Stelco, needs for more land were met by filling in adjacent parts of the harbour.

Immediately inland from the spreading industrial waterfront, on slightly higher ground, roughly above the 82.5 m (275 ft) contour, the land was used mainly for working-class homes and was the object of intense speculation. From 1906 to 1915, an average of nineteen new surveys per year was registered in the city as compared with six per year in the previous thirty years and only 1.36 per year in the following quarter-century. But the area available for residential use far exceeded the effective demand, leading to a patchwork settlement pattern. Urban facilities could not keep pace with the easterly expansion, except for streetcar service, which in some places actually preceded development, encouraging the sprawl. Contemporary accounts speak of isolated groups of buildings rising above the 'eternal mud.' In 1906 it was reported that, to get water, some local residents travelled up to half a mile to a spring at the corner of Ottawa and Barton streets (Campbell, 1966).

To the west, urban expansion was held back by the presence of the Chedoke Valley. Although not particularly deep, the valley had been used for decades as a garbage dump and was thus both a psychological and a topographic barrier. With no urban services, the Westdale Terrace remained essentially rural except for a few shacks overlooking the dump.

However, by 1914 the residential potential of this area was recognized; in that year it was annexed by the city and provided with good access by the construction of the McKittrick Bridge on King Street.

Southward, the escarpment was still an uncompromising obstacle and one which the city was not yet ready to assault. East of Wentworth Street as far as Kenilworth, its gradient was indeed so steep, as a result of undercutting by the waves of Lake Iroquois and some quarrying, as to preclude any use save for a roadway which clung to the rock wall. Furthermore, in this stretch, the TH&B line running at the base of the slope constituted an additional barrier. To the west, however, and particularly west of James Street where the Iroquois Bar had protected the escarpment from wave action, the ascent was more gradual. Here upper-class homes continued to be built.

A relatively small area was involved because the more wealthy group among the population had not increased in numbers as rapidly as the workers, particularly because the decline of Hamilton's commercial function had resulted in the departure of a number of offices to Toronto and elsewhere. Nevertheless enough well-to-do families remained to support

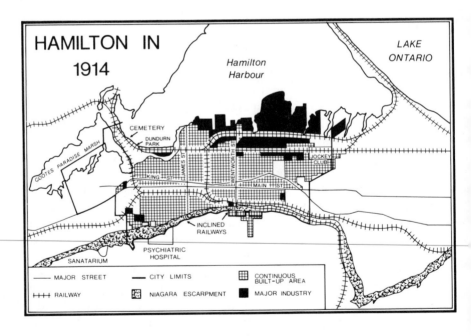

Figure 8.3

a golf club above the Chedoke Valley, a yacht club and luxurious summer hotel on the beach strip, and a race track, known as the Jockey Club, which extended from Ottawa Street all the way to Kenilworth, north of Barton.

The Mountain area still remained separate from the city, though its population was slowly increasing. To serve the residents, a second inclined railway was constructed in 1900, at Wentworth Street, and a water tower was built in 1913 (City of Hamilton, 1926). But the only annexations were of two small parcels of land on the brow, one of which was later used as the site of a hospital.

The economic, demographic, and geographic growth of Hamilton was supplemented by considerable institutional development. In 1900, a parks board was set up, and by 1905 the city had acquired Victoria Park, Dundurn Park, and Mountain Drive Park. Also in 1905, the Hamilton Health Association was founded, taking immediate action to deal with a major problem of the working classes through the opening of a tuberculosis sanatorium, located appropriately on the Mountain, away from the congestion and air pollution of the city. In 1911, Hamilton obtained its public library. In 1912 the Hamilton Harbour Commission was established, one of its first projects being to reclaim some land at the foot of Mary Street for a small park. In 1914, the Hamilton Hydro-Electric Commission was formed, street lighting was taken over by the city, and the power-distribution network was extended to cover all built-up areas. In that year alone, the number of consumers increased from 6,500 to 10,000.

Thus, by 1914 Hamilton had become an industrial city, elongated in an east-west direction in response to local topography, and with a rudimentary spatial differentiation of its functions. In some ways it was a prosperous city, because, despite a business slowdown towards the end of the period, much wealth had been generated. But most of the city's population were of the working class, and the workers remained poor, ill-housed, and unhealthy. Economically, much had been achieved; socially, conditions were little better than they had been a quarter of a century earlier (City of Hamilton, 1945).

UPS AND DOWNS, 1914-38

The effect on Hamilton of the First World War was mixed. On the one hand, certain industries, notably those involving iron and steel, expanded greatly, with capacities pushed to the limit despite the addition of new facilities. An important manifestation of this expansion was the creation in 1917 of the Dominion Foundaries and Steel Company (Dofasco) by the amalgamation of three previously existing companies. In 1914, these companies employed a total of 190 workers. By 1918, Dofasco had a work force of 2,283. During the war Hamilton's population passed the 100,000 mark.

However, the city was ill-equipped to deal with the influx of workers. The already inadequate stock of housing was subjected to an even greater strain. In and around the industrial zone, a number of boarding houses sprang up. Here conditions were particularly depressing; one such house was found to be inhabited by seventy-five people, sharing a single bathroom. Various investigations into social conditions were undertaken but nothing concrete was done by the city government, in which no single body was responsible for housing. The only public facility added during the war was Gage Park. This was, indeed, a valuable recreational area, but on balance the war did not make Hamilton a better place in which to live, at least for the poor.

The war did not have much influence on the general pattern of urban development. Nor, strangely enough, did the adoption of motor vehicles. One of the main obstacles to their use for regional transportation continued to be the lack of suitable roads. Not until 1916 was there a paved highway to Toronto. Another five years elapsed before the road to Dundas and Waterloo was paved. Mass transit by motor vehicle did not become common for nearly a quarter of a century after the arrival of the first automobile. Bus service to Toronto started in 1920, and not until 1923 was a comprehensive network of bus routes established to link Hamilton with all nearby communities. Even so, buses only gradually replaced the electric radial cars, the last of which ceased operations in 1931. The shift from one type of carrier to another was not sufficiently dramatic to induce any important population shifts, except on the Mountain.

Because of the height and steepness of the escarpment, the Mountain area had never been served either by streetcars or by radial cars, while the inclined railways, though useful, covered only a few hundred feet of horizontal distance. For the level land above the escarpment, public transportation came into existence in 1923, when bus connections with the city were established. The results, in terms of the people attracted to the Mountain, were impressive.

Yet, housing conditions on the Mountain were in some ways even worse than in the poorest parts of the lower city. Not only did the area lack such basic services as fire and police protection, but there were no sewers, and piped water reached only a small part of the population. The only sanitary facilities were outhouses, but, because the limited absorptive capacity of the thin layer of earth over the bedrock, they were supposed to be equipped with tin-lined containers which were periodically pumped out by private contractors. Some were so equipped; many were not, and the groundwater was generally severely polluted, a grave situation in an area where most of the domestic water supply came from wells. An article in the Hamilton *Spectator* described the conditions as 'deplorable and unprecedented,' but not until 1927, following an outbreak of scarlet fever, did city officials move to correct this situation. In 1929, the city annexed the main populated area north of Fennell Avenue. Following annexation, improvements were made to local services, including bus transportation. In 1932, though not usable by buses, a new Mountain access route was constructed starting at Kenilworth Street and reaching the top through what is known as the Sherman Cut. These improvements led to the closing of the two inclined railways, James Street in 1931 and Wentworth Street in 1936. By 1935 the Mountain area had a population of 10,000 while the city as a whole had just over 150,000 people.

To the west, some expansion took place during the war as various industries were established in the Chedoke Valley near the TH&B railway lines, and some workers' homes were built on adjacent parts of the West-dale Terrace. Development in this direction began in earnest in 1923. In that year the city decided to donate a parcel of land on the western margin of the terrace to McMaster University so that it might move to Hamilton from Toronto. The effect on local land use was immediate and profound. In the certainty of rapid high-class residential development, Westdale was laid out as a planned community, centred on the terminus of the streetcar line, which went into operation in 1924. Before long, development spread to the area south of Main Street, which was annexed in 1929. In 1930, McMaster opened its doors. Meanwhile, in 1926 another important initiative was under way with the establishment of the Royal Botanical Gardens. Starting with only 40 ha (100 ac), the Gardens rapidly expanded to include much of the land adjacent to the McMaster campus and also parts of the Iroquois Bar. In the latter area, the cemetery was becoming overcrowded, and to obtain space for future burials, the city, in 1921, annexed the tip of the bar's northern arm.

Throughout this period, urban expansion to the east, in the direction of least physical difficulty, was slow, even though the city was making a special effort to improve the condition of the main streets. The only annexation for residential use was a relatively small area to the south, between Ottawa and Kenilworth streets. The need for land was mitigated, to some degree, by the availability of more attractive residential areas to the south and west. But the main reason for slow eastward expansion was the existence within the city limits of hundreds of vacant lots, relics of the pre-war speculation fever. Growth was thus more by the infilling of the existing urban area than by its lateral extension. Consolidation was also achieved through the construction of many multi-family buildings. Between 1921 and 1931, the proportion of Hamilton households living in apartments increased from 4 per cent to over 15 per cent.

The East End did benefit, however, from the British Empire Games, held in Hamilton in 1930. Facilities constructed for the games and subsequently open to the general public included a swimming pool and an athletic field, now known as Ivor Wynne Stadium. Meanwhile, outside the city limits, a small airport had been set up in 1929 near the Red Hill Creek. Established for their use by members of the Hamilton Aero Club, founded the previous year, the airport was essentially another sporting facility for the elite, but it did provide some general social benefits as it permitted daily air-mail service to be brought to the city.

To accommodate new industry, in 1920 a strip of land was annexed extending the city limits from Kenilworth Street to Stratherne, but in general previously existing industry obtained such new land as was required by further reclamation of swamps and parts of the harbour. Some shoreline modifications were also necessary to accommodate the large ships which began to come to Hamilton after 1932. In that year, a completely new Welland Canal was opened, capable of accommodating vessels up to 225 m (750 ft) in length with cargo capacities of 25,000 tonnes (28,000 t), ten times the capacities handled by the previous canal. For the first time, the large bulk carriers, which for years had been plying the waters of the upper Great Lakes, could enter Lake Ontario, where one of the most important destinations, particularly for cargoes of iron ore, was Hamilton.

However, such progress as was being made in Hamilton during the 1920s and immediately thereafter was brought to an abrupt halt by the onslaught of the Great Depression. Because its industries depended mainly on national and international markets, Hamilton suffered more severely during the depression than did other cities supported to a greater degree by trade with local hinterlands. Many small industries failed, and even a giant such as Dofasco wavered on the edge of bankruptcy. In March 1933, 22.5 per cent of all families in the city were on relief. Desperate men roamed the streets, knocking on doors to beg for food. Many left; from 1931 to 1936 the city population actually declined by 1.4 per cent, the first intercensal decline since 1857-61. The civic treasury was so depleted that the new Mountain Hospital, completed in 1932, could not be opened for six years for lack of furnishings and equipment. The only positive elements of the urban landscape emerging from these dark days were two make-work projects: the Rock Gardens, built in an abandoned gravel pit on the Iroquois Bar, and the Sunken Gardens, on land now occupied by Mc-Master's Health Sciences Centre.

Elsewhere, urban development was less kind to the environment. Garbage continued to be dumped into the Chedoke Valley, while new dumps were opened on the Mountain and in the East End at the swampy mouth of the Red Hill Creek. In 1925, a proposal that the city build an incinerator was turned down on the grounds that when existing dumps were full, the entire Dundas marsh was available for garbage disposal. A few years later, an incinerator was in fact constructed, but it could only handle a fraction of the garbage; the three dump sites remained in use.

Yet even more environmentally damaging than the method of handling garbage was the city's sewage-disposal system. As long as sewers had existed in Hamilton, they had discharged their effluent directly into the harbour with no treatment whatsoever. As the population grew, and as domestic wastes were supplemented by ever-increasing outflows of chemically active fluids from industry, the harbour became more and more polluted. It was first reported unsafe for swimming in 1914, but not until 1923 was the water officially declared to be polluted, bringing to an end the cutting of ice in winter for domestic use in summer. When a sewage-treatment plant was built, its only function was to break up the larger solid particles contained within a small fraction of the total sewage flow. The problem seems not to have been taken very seriously at the time. As late as 1927, the public were still using a 'bathing beach' at the foot of Wentworth Street, despite the presence of nearby sewage outlets both to the east and to the west. And, as nothing was done to alter the situation, water quality continued to deteriorate. By 1947, the coliform count in the harbour was seven times what it had been in 1923. By 1958 it was thirty times as high.

In summary, the period 1914-38 was one in which Hamilton went ahead in some aspects and backward in others. There had been some relatively good years, particularly at the end of the 1920s, and some parts of the city, notably the Westdale area, had made impressive advances. But as 1938 drew to a close, the city exhibited greater contrasts between rich and poor, and faced problems of greater magnitude than ever before in its history.

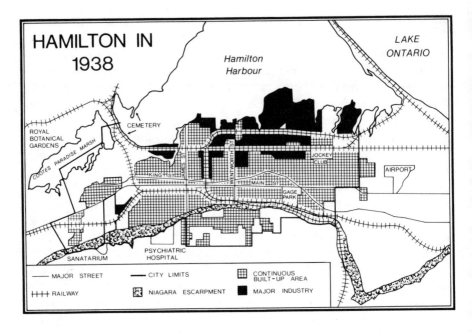

Figure 8.4

TAKING STOCK, 1938-50

The impact on Hamilton of the Second World War was much greater than that of the previous conflict. Not only did the war last longer, but it made far greater use of heavy weapons and equipment. The contrast is illustrated by the experience of Dofasco. In the Second World War the company made such items as gun barrels and armour plate for ships and tanks. During the First World War, the main items of military production were stirrups and other bits of hardware for the cavalry.

Of course, the initiation of hostilities brought an end to the depression. Industries returned to full capacity and then built new facilities, including a 63.5 tonne (70 t) blast furnace, largest in the British Empire, inaugurated by Stelco in 1943.

The demand for labour in the war industries stimulated a massive inflow of workers. In 1941, 74 per cent of all gainfully employed persons in the city were in the manufacturing sector, more than half of them in metal-working factories. Yet, although the new arrivals could obtain jobs they could not easily find places to live. Following years of economic stagnation when few homes were built, the sudden influx of labourers put unprecedented pressure on the city's stock of housing. In 1942, in the central part of the city there were, on the average, three families for each two residential units and five persons for each two beds. Flies and rats constituted serious problems, as did air pollution, not only from industrial fumes but also from domestic chimneys, since 95 per cent of all homes were heated by either coal or coke. In 1944, the Hamilton Health Office described the housing situation as 'deplorable,' attributing the spread of various infections diseases to the overcrowded and unsanitary conditions.

As an emergency measure, it was decided to provide temporary accommodation in the form of small prefabricated cottages without basements, together with a number of larger barrack-type buildings for single persons and/or multi-family use. The latter were in the northeast, close to or within the industrial area. The former were more widely scattered, though a major concentration was in the East End and another on the Mountain. In the latter area were built 581 'wartime houses,' 34 per cent of the city's total. By 1945, one-third of the entire Mountain population was living in this wartime housing, which turned out to be much less temporary than had been initially indicated. All these units were still in place by 1950, and though a few had been made 'permanent' by the addition of basements and other forms of upgrading, on balance they were rather unattractive parts of the urban landscape.

It is enlightening to examine a map of housing conditions produced in 1945 by a firm of planning consultants. Although only a handful of blocks were classified as 'slums,' all residential areas north of Barton Street were considered to be 'blighted' as was also the entire area between Barton and King from Wentworth Street in the east to Locke Street in the west. In addition, more than half of the remaining housing below the escarpment and almost the entire residential area on the Mountain were classed as 'declining.' It was noted that piped water and/or sewers were absent in a number of these areas including many blocks on the Mountain, much of West Hamilton, and even the earliest settled parts of Westdale.

According to the consultants, only one-sixth of the population lived in 'sound' neighbourhoods. These included the traditional upper-class district near the base of the escarpment west of James Street and two more recently built-up areas, one in Westdale and another around Gage Park to the east. It was suggested that much deterioration in housing quality was the result of the desire of members of the middle class to escape from the central city. When they could, they moved to the better new subdivisions or to the suburbs, but they left behind residences in decay for lack of maintenance and architecturally unsuited to the needs of the poorer people who took them over.

The growing residential separation of different income groups was accompanied to some extent by a segregation based on country of origin as migrants from places other than the United Kingdom tended to form their own neighbourhoods. Thus, the working-class parts of town began to be divided by language, and to some extent by religion, into a number of fairly distinct and visibly identifiable ethnic districts.

Ethnic diversity, however, did not prevent the workers, facing common difficulties in daily living, from joining together in a new militancy. They had acquired new bargaining power during the wartime period of labour shortages, and used this power to bring about a large-scale unionization, which employers were forced to accept. Once the war ended, better conditions were demanded in sometimes bitter confrontation, the most memorable being the great Stelco strike of 1946 which lasted 81 days and was accompanied by much violence.

Though disruptive at the time, these and other disputes did bring higher wages and other benefits to the working classes, making it possible for them to improve their living standards. Nor does it seem that the companies suffered greatly. Despite concessions granted as a result of the strike, Stelco was able, in 1948 and 1949, to undertake a major program of expansion, including a huge land reclamation in the harbour. Business generally prospered as the increasing affluence of the working class provided the principal fuel for a great post-war building boom. From 1946 to 1951, 4,500 detached dwellings were constructed, most of them comfortable bungalows in the East End or on the Mountain.

To accommodate this new development, Hamilton's areal growth during this period was mainly to the east. First, in 1943, the city annexed a strip of land between Main Street and the harbour, pushing the municipal boundary as far as Parkdale Avenue. Then, post-war growth led, in 1949, to the largest annexation in the history of the city to that date. Acquired for both residential and industrial purposes was an extensive area, stretching eastward to the banks of the Red Hill Creek. At the same time, the hilly upper reaches of the Red Hill Valley were annexed for future development as a city park.

Also annexed in 1949 were two relatively small areas in the west. One included part of the McMaster campus and a small residential zone farther west. The other was a previously ignored part of the Chedoke Valley into which industry was spreading from the previously established industrial concentration by the TH&B freight yard.

Yet the most significant annexation of 1949 was on the East Mountain, the first indication that the city recognized that its future expansion would have to be southward. The change in outlook did not come easily, because of the high costs involved. Sewers were particularly expensive to construct as they had to be laid in trenches excavated in bedrock. And the difficulty of maintaining all-season road access had been highlighted by the great snow of 1944 which almost completely isolated the Mountain for more than a week. Still, the decision was made to achieve the necessary integration, and, as the period ended, a major step forward was taken with the opening of the new Jolley Cut, a four-lane highway with moderate grades, providing, for the first time, a modern road link between the upper and lower parts of the city. Between 1946 and 1951, the population of the Mountain area doubled.

At the same time, this new attitude with respect to the Mountain was only one indication that the management of urban development was at last perceived as necessary. Though town planning had been discussed as early as 1912, not until the late 1940s did it become a reality. Only then did the city demonstrate its seriousness of purpose by turning for

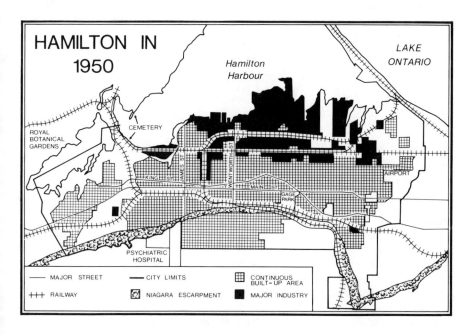

Figure 8.5

assistance to aerial photography, a type of technology which, though not particularly new, was the only practical method of achieving a comprehensive land inventory. In 1950, the task was not yet completed but was well on its way. Every parcel of land and every structure were being identified and mapped at the scale of one hundred feet to the inch. For the first time, civic authorities began to be fully informed and really concerned about land use, not only in the hundreds of previously unregistered and unassessed parcels of land turned up by the survey, but also in such major land-use anomalies as the Jockey Club and the east-end airport.

Concern over land use was accompanied by an awareness of the problems of traffic within the city. Even the desirability of installing parking meters had not been studied before 1945, and growing congestion, caused by an increase in the number of vehicles – an increase as much as 30 per cent between 1945 and 1947 alone – led to the discussion of various schemes to improve circulation.

Thus, as the period ends, Hamilton had reached a new stage of maturity. It had finally come to terms with the constraints of its site, and it had at last assumed responsibility for the details of its future growth (City of Hamilton, 1958). Many problems remained to be dealt with, but the fact that the difficulties had become the objects of careful analysis gave promise that they could eventually be solved.

REFERENCES

Campbell, M.F. 1966. *A Mountain and a City: The Story of Hamilton.* McClelland and Stewart, Toronto

Census of Canada 1911. King's Printer, Ottawa

City of Hamilton. 1926. Annual Report of the City Engineer

– 1932. Industrial Department Promotional Brochure

– 1945. City Planning Committee, Survey of Existing Conditions

– 1958. Urban Renewal Study

Hamilton *Spectator* 25 October 1984. *Hamilton, Our Lives and Times.* Special issue

Johnston, C.M. 1958. *The Head of the Lake: A History of Wentworth County.* Robert Duncan and Company, Hamilton

Middleton, D.J., and Walker, D.T. 1980. Manufacturers and industrial development policy in Hamilton, 1890-1910. *Urban History Review*, February, 8: 20-46

Petrie, F.J. 1967. Canal development in the Niagara Peninsula. In St Lawrence Seaway Authority (1967)

St Lawrence Seaway Authority. 1967. *Welland Canal Relocation.* Ottawa

Walker, D. 1980. *The Human Dimension in Industrial Development.* University of Waterloo, Department of Geography Publication no. 16

Weaver, J.C. 1982. *Hamilton: An Illustrated History*. James Lorimer, Toronto

Westland, S.I. 1951. Land Transport Geography of Hamilton. MA thesis, Department of Geography, McMaster University

Wingfield, A.H., ed. 1946. *Hamilton Centennial 1846-1946: One Hundred Years of Progress*. Davis-Lisson Limited, Hamilton

Wood, H.A. 1960. *The Site of Hamilton and Its Influence on the Development of the City*. Canadian Association of Geographers, Education Committee, Bulletin no. 7

CHAPTER 9

Social change in Hamilton, 1961-1981

S.M. TAYLOR

The social geography of the city has been aptly described as the 'urban mosaic' (Timms, 1971) with variations in culture, social group, life cycle, and other characteristics of the urban population providing the colour and texture. The patterning is by no means random. Regularities underlie the particular features of individual cities. While Hamilton has a social geography which distinguishes it from Toronto or Montreal or Vancouver or, for that matter, from Los Angeles or Paris, it shares some common features with other cities in how different social, cultural, and demographic groups are spatially distributed. The challenge for the urban analyst is to do justice to the blend of particular and common features. A social geography which fails to convey something of the personality of place in emphasizing only the abstractions which pertain to most, if not all, places is seriously lacking. But equally, single-city descriptions, largely ignoring structural elements and processes, which are key to understanding the factors shaping cities in general, are similarly guilty of providing only part of the story.

In a selection of essays like this one, which focuses on a single city, chapters vary in their relative emphasis on the general and the particular. Other essays have described different historical and contemporary features of the social geography of the Hamilton region. This essay is designed to complement them by focusing on the more general patterns which characterize the residential areas of the city, patterns which to some degree Hamilton shares with other cities. Specifically, the essay will describe and analyse how these more general characteristics of the social geography change over time in response to the combination of social, economic, and political factors which together ensure that the city is always in a state of social and spatial transition, small though the changes may appear in the short run.

Given this focus, the immediate issues are to determine what social-geographic characteristics to use as indicators of change, and, second, to select an appropriate method for measuring and analysing the direction and magnitude of changes in different parts of the city. These issues are dealt with in the following section as a prelude to the analysis which uses census data to assess social change in Hamilton during the period 1961-81. The final section summarizes the conclusions of the analysis and attempts to link them with relevant material from other essays.

THE SOCIAL GEOGRAPHY OF THE CITY

Casual observation of any cityscape quickly reveals a diverse social geography. For example, an east-west transect of Hamilton might begin at the city boundary with Stoney Creek in an area of fairly new residential development and a mix of single-family, townhouse, and apartment dwellings. Moving west along Barton Street reveals a succession of older, industrial neighbourhoods, each with distinctive characteristics and bearing the imprint of the cultural and ethnic backgrounds of the residents, including Ukrainian, Italian, and Portuguese immigrants. Beyond James Street, to the west of the downtown core, the scene changes again and neighbourhoods in various stages of transition are encountered. Middle- and high-income apartment and condominium buildings are juxtaposed with more modest, older, single-family housing in areas such as the Durand neighbourhood. It is in this general area, too, that gentrification is most obvious, with the interior and exterior refurbishing of turn-of-the-century structures. Highway 403 divides the inner-city neighbourhoods from the suburban areas to the west, including the 1930s planned suburb of Westdale which takes in McMaster University. Still farther west, approaching the boundary with Dundas, inter-war single-family homes and more recent apartment buildings characterize the west-end neighbourhoods.

This type of very general description gives some clues to the characteristics which might be used to classify social areas and to describe the social geography of the city. Among these are the type and age of the housing stock, the economic status of the population, the demographic composition of households, and the cultural background of residents. To some degree, casual observation can provide a basis for delineating and distinguishing neighbourhoods, and for many of us our image of the social geography of the city is constructed in this way. The academic community, primarily urban geographers and sociologists, have devised more formal methods for classifying the social areas of cities. The different approaches which have emerged have two tasks in common. The first is the identification of key variables for classifying areas; the second involves mapping the areas once they are classified and interpreting their spatial distribution.

No attempt is made here to review these various methods. The interested reader is referred to a suitable text dealing with the topic (Herbert, 1971; Timms, 1971; Ley, 1983). For present purposes, it is sufficient to note that commentators typically distinguish four main approaches to the definition of social areas: morphological analysis, the natural-area approach, social-area analysis, and factorial ecology. The first two approaches depend primarily on field observation to classify areas in terms of their combined physical and socio-cultural characteristics. The other two approaches are based on the analysis of census data.

The work of the social-area analysts is important in identifying general dimensions of social differentiation which might serve as a basis for classifying urban areas. Shevky and Bell (1955) argued that the social areas of cities reflect very general trends within society at large and attempted to integrate these trends in their construction of social-area analysis. They identified three trends as particularly important: increasing occupational specialization, changes in family structure, and the separation or segregation of ethnic and other lifestyle groups. These were operationally defined as three constructs (social rank or economic status, urbanization or family status, segregation or ethnic status) which could be measured using census variables. Shevky and Bell applied this method to classify social areas in the San Francisco Bay region and were able to show, by mapping the results, systematic patterns in the spatial distribution. The three constructs developed by the social-area analysts re-emerge as fundamental dimensions in studies based on the factorial-ecology approach.

Like social-area analysis, factorial ecology uses the definition of dimensions of social differentiation as the basis for classifying and mapping social areas. However, in factorial ecology the dimensions are not prescribed a priori on some theoretical basis, but rather are empirically constructed from a multivariate analysis of census data. The advent of factorial ecology coincided with the availability of high-speed computers required to perform complex statistical analyses of large data sets (e.g., census data).

HAMILTON'S SOCIAL STRUCTURE, 1961-81

Dimensions of Measuring Social Change

Factorial ecology involves a statistical analysis of census data with the objective of defining dimensions of social structure which can be used to classify census areas in terms of social composition. The statistical procedure used is factor analysis. This procedure summarizes correlations among census variables. Subsets of highly interrelated variables define underlying factors which can be interpreted as dimensions of social structure. For example, various census measures provide indices of the economic status of an area, including average income levels, unemployment rates, and percentages employed in certain occupational categories. Such variables are typically strongly intercorrelated and combine in a factor analysis to define a single factor which might be labelled socio-economic status. Similarly, other subsets of interrelated census variables combine to define other factors. In this way, a relatively small number of social dimensions can be defined to provide a parsimonious description of social structure.

As well as defining dimensions of social structure, factor analysis calculates scores for each census area (tract) on each dimension which can be used to map the spatial distribution of social characteristics. Maps can be compiled based on area scores for a single factor or a combination of factors if a composite typology is desired.

Factorial ecology is the basis for this analysis of social change in Hamilton for the period 1961-81. A separate factor analysis was performed for each of the three census years 1961, 1971, and 1981. To facilitate comparing results over time, the same census variables were selected for inclusion in each analysis. The twelve variables selected (table 9.1) represent demographic, economic, and cultural characteristics and include variables shown to be key indicators of socio-economic, family, and ethnic status in previous studies. These variables included: average number of children per family (CHLDPF); median family income (MFAMINC); proportion male (PRMALE); proportion over 65 (PRGT65); proportion single over 15 (PRSGT15); proportion male unemployed (PRMUNPL); proportion female unemployed (PRFUMPL); proportion male in professional occupations (PRMPROF); proportion born in Canada (PRBNC); proportion of European origin (PREURO); proportion with less than Grade 9 education (PRCTG9); and proportion with university education (PRUNI). The census tracts for the City of Hamilton comprised the units of observation and analysis. As a result of increases in population, the number of census tracts increased from 65 in 1961 to 93 in 1981. This increase and associated changes in some tract boundaries cause some complications for areal comparisons over time.

As already noted, most factorial-ecology studies have shown economic status, family status, and ethnic status as major dimensions of social

TABLE 9.1
Factor analysis results, 1961-81

Variable	1961			1971				1981			
	Factor 1 (economic)	Factor 2 (family)	Factor 3 (ethnic)	Factor 1 (SES)	Factor 2 (family)	Factor 3 (ENDER)	Factor 4 (ethnic)	Factor 1 (SES)	Factor 2 (family)	Factor 3 (ethnic)	Factor 4 (OCCUP)
CHLDPF	-0.13	0.82	0.05	0.15	0.89	-0.13	0.10	0.07	-0.48	0.15	0.65
MFAMINC	0.81	0.16	-0.40	-0.91	0.30	-0.10	-0.07	-0.64	-0.27	-0.00	-0.05
PRMALE	-0.32	0.56	0.63	0.22	0.23	-0.67	0.16	0.14	-0.02	-0.03	0.71
PRGT65	0.03	-0.82	-0.03	0.18	-0.73	0.41	-0.07	0.12	0.35	-0.00	-0.28
PRSGT15	0.07	-0.81	0.06	-0.05	-0.11	0.92	-0.02	0.04	0.86	0.08	-0.07
PRMUNPL	-0.51	-0.20	0.65	0.78	-0.15	-0.00	-0.02	0.70	0.25	0.15	0.06
PRFUMPL	-0.32	0.16	0.46	0.53	0.06	-0.29	0.28	0.27	-0.08	-0.06	0.03
PRMPROF	0.98	0.01	-0.15	-0.81	-0.11	0.37	-0.17	-0.48	0.34	0.04	-0.33
PRBRNC	0.32	0.68	-0.42	-0.34	0.50	0.03	-0.60	-0.08	-0.11	-0.65	0.10
PREURO	-0.11	-0.17	0.77	0.09	0.16	-0.09	0.63	0.00	-0.10	0.97	0.20
PRCTG9	-0.55	0.09	0.30	0.84	0.20	-0.14	0.39	0.84	0.01	0.39	-0.13
PRUNI	0.90	-0.36	-0.13	-0.67	-0.43	0.11	-0.12	-0.64	0.52	-0.01	-0.27

structure. Economic status is defined by income, education, and occupation variables. Family status combines household size, marital status, and number and ages of children. Ethnic status brings together measures of minority-group representation.

The results from the factor analyses of the census data for Hamilton for 1961, 1971, and 1981 (table 9.1) reveal the same three dimensions. For 1961, three factors were defined: economic status (factor 1), family status (factor 2), and ethnic status (factor 3). The 1971 analysis yielded four factors, the additional one (factor 4) being labelled 'gender status.' Four factors were defined in the 1981 analysis also. In this case, the extra dimension (factor 4) was most strongly related to occupational status. The labelling of dimensions depends on the variables most strongly related to each factor (i.e., the variables having the highest loadings, whether positive or negative). For example, the first dimension from the 1961 analysis is labelled 'economic status' on the grounds that the highest-loading variables are median family income (positive) and male unemployment (negative).

The consistency of the results for the three years is important for the subsequent analysis involving comparisons between years in socio-spatial structure based on the three common dimensions: economic, family, and ethnic status. The comparisons are only valid if the composition of the factors is consistent across the three census years. Consistency was tested in this case by a correlation analysis using the variable loadings. The results confirmed high correlation in the composition of each factor over time.

Spatial Patterns of Social Characteristics

The scores for each census tract on each of the three common dimensions of social structure are the basis for examining the spatial distribution of social characteristics. Maps (figures 9.1-9.9) were compiled for each dimension for each census year. These allow a description of spatial patterns for each year and of changes over the twenty-year period on each dimension. The maps are most usefully summarized by considering each dimension in turn.

Considering first the spatial pattern for *socio-economic status* (figures 9.1-9.3), there are some obvious consistencies over the 1961-81 period. The west end of the city below the escarpment has maintained a high socio-economic status throughout the period. Equally, the north and east ends have remained relatively low-status areas. The area above the es-

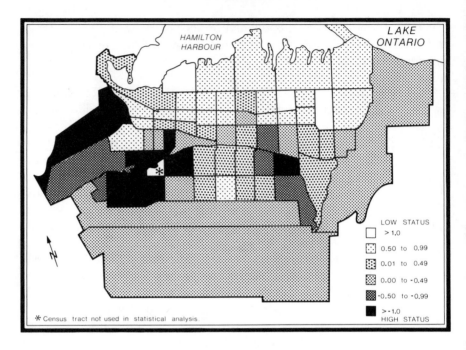

Figure 9.1 Spatial pattern for socio-economic status in Hamilton, 1961

Figure 9.2 (*right, top*) Spatial pattern for socio-economic status, 1971

Figure 9.3 (*right, bottom*) Spatial pattern for socio-economic status, 1981

carpment exhibits the clearest changes linked to the residential growth that has occurred there over the last twenty years. The extent of the growth is evident from the increase in the number of census tracts. Notice that in 1961 there were only two census tracts south of Mohawk Road. By 1971 this had increased to five and by 1981 to fifteen (figures 9.2 and 9.3). Initially, this was an area of slightly above-average economic status. Residential development in the decade 1961-71 resulted in an increase in economic status in the western and northern sections and in the following decade this increase spread to the eastern section as well, such that by 1981 most of the South Mountain area was well above average status. In general, therefore, the major change over the two decades has been the eastward and southward extension of higher-status districts previously

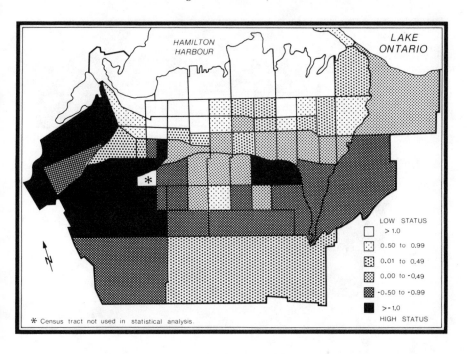

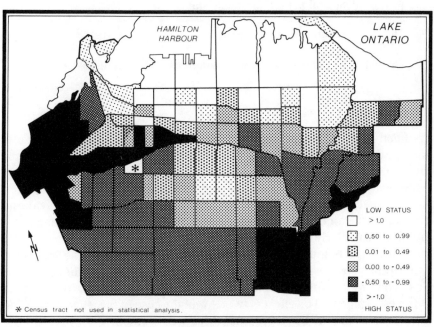

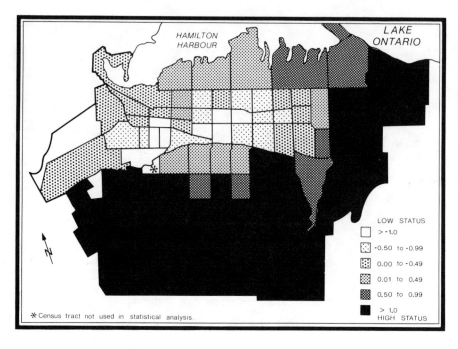

Figure 9.4 Spatial pattern for family status, 1961

Figure 9.5 (*right, top*) Spatial pattern for family status, 1971

Figure 9.6 (*right, bottom*) Spatial pattern for family status, 1981

confined primarily to the west end. This change in large measure reflects the spatial distribution of new single-family housing construction in the city.

A consistent finding of many previous factorial-ecology studies has been that socio-economic status tends to conform to a sectoral pattern with the high-status sector of the city located away from areas of low environmental quality, particularly areas of industrial concentration. The predominantly low-status tracts found in the north and east and close to the industrial waterfront provide some support for the same trend in Hamilton, but close conformity with a simple sectoral pattern is prevented by the effect of the escarpment. The effect is twofold. In the case of the tracts bordering the eastern edge of the escarpment the scenic amenity afforded by the

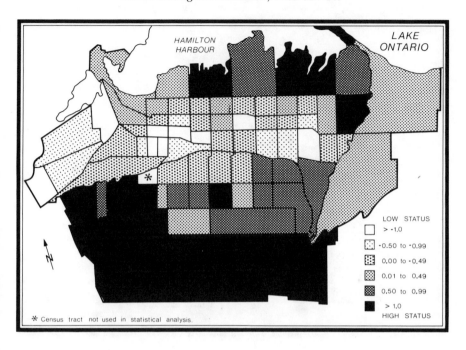

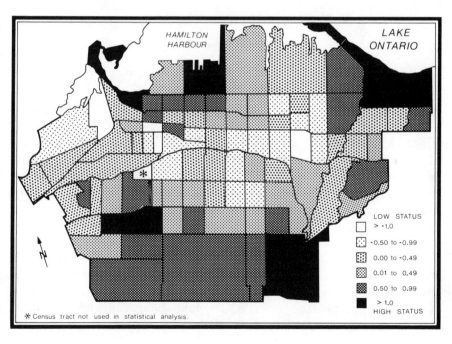

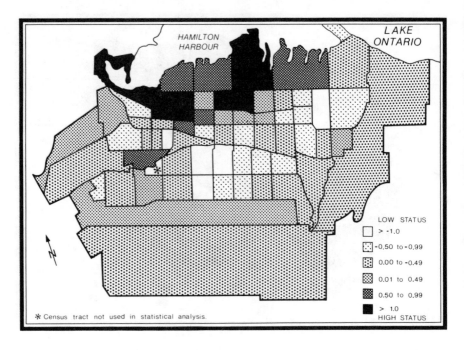

Figure 9.7 Spatial pattern for ethnic status, 1961

Figure 9.8 (*right, top*) Spatial pattern for ethnic status, 1971

Figure 9.9 (*right, bottom*) Spatial pattern for ethnic status, 1981

elevation and the view east towards the Niagara Peninsula has countered the relative proximity of the industrial area such that these tracts have maintained a relatively high status. More generally, the escarpment has served to increase the effective distance between the mountain tracts and the lower-status north and east end.

In many cities, *family status* exhibits a zonal pattern with status increasing with distance away from the central business district (CBD) and inner city, reflecting a general preference among households with children for suburban environments. This pattern is supported for Hamilton (figures 9.4-9.6) although the maps suggest that the clear distinction between a low–family-status inner city (i.e., few households with children) and a high–family-status suburban area has been weakening somewhat over

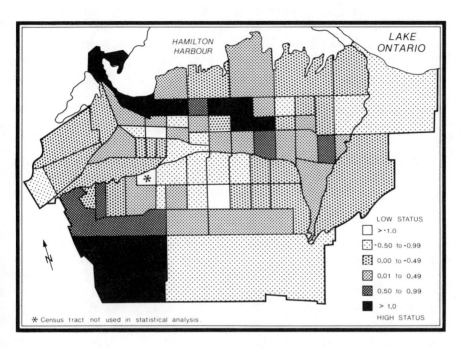

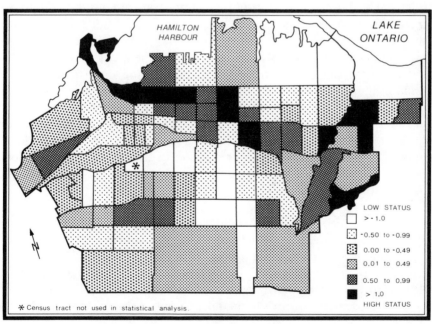

time. The zonal pattern which is so clear in the map for 1961 (figure 9.4) is not nearly as obvious in the 1981 map (figure 9.6). Particularly noticeable is the decline in family status for several of the tracts of older housing in the northern section above the escarpment. Equally, the evidence of increasing status for some inner-city tracts runs counter to a simple zonal arrangement. This latter trend is probably linked to foreign (especially South East Asian and Portuguese) immigration flows to the inner city during the 1970s. The low family status for the west-end tracts throughout the period reflects the fact that these areas have a relatively high proportion of elderly households. If the tracts in Dundas and Ancaster had been included, the suburban high–family-status zone would have extended to encompass the western as well as the southern and eastern margins of the city.

In the case of *ethnic status* (figures 9.7-9.9), the inner city has traditionally emerged as the primary area of concentration of foreign immigrants. The accepted explanation for this concentration is that the inner city affords cheap housing commensurate with the means of low-income immigrant group and access to low-paying jobs. The map of ethnic-status distribution for 1961 (figure 9.7) shows that such a concentration existed in Hamilton at that time with the areas of highest status in the north end of the inner city and extending eastwards along Barton Street. In 1971 (figure 9.8) those same areas of high status remain with the addition of tracts on the West Mountain, an area of new residential development in the preceding decade to which many second- and later-generation immigrants were attracted. By 1981 (figure 9.9), the areas of concentration are more dispersed. The high-status tracts in the north and east end remain; the concentration on the West Mountain is no longer evident as a result of the mix of social groups attracted to the new housing constructed in that area in the 1970s; new areas of high ethnic status are found on the eastern margins of the city, bordering Stoney Creek. Notice that these maps show the distribution of ethnic groups in general; the concentrations of specific groups and their movements over time cannot be determined directly from these results.

Social Change in the Inner City

We turn from the description of city-wide social patterns to a more detailed examination of changes in the inner city in the period 1961-81. The inner city is an appropriate focus of attention because it is there that the urban-development trends outlined in the earlier part of this essay have in many cases had their most obvious manifestations. Suburbanization, for ex-

TABLE 9.2
Average social dimension scores for inner city tracts,
1961-81

Social dimension	1961	1971	1981
Socio-economic status	0.18	0.59	0.41
Family status	− 0.68	− 0.70	− 0.37
Ethnic status	0.46	0.01	0.27

ample, contributed directly to the social and economic decline of the inner city. Subsequently, urban renewal, gentrification, and the so-called 'back to the city' movement gave impetus to the revitalization of inner-city neighbourhoods. The question we consider here is whether there is any evidence of the effects of these general trends in terms of social changes in inner-city Hamilton.

The inner city was defined as the tracts including and surrounding the central business district, extending west to Highway 403, north to the harbour, east to Wentworth Street, and south to the escarpment. This area comprises eighteen census tracts whose boundaries have changed very little over the twenty year-period, allowing valid comparison of social characteristics over time. The census-tract boundaries for 1981 are shown in figure 9.10. The analysis of social change was approached in two ways: by comparing the scores for each year on each of the three dimensions of social structure averaged over all eighteen tracts, and by identifying specific tracts in which major changes had occurred between census years.

The average scores on the three dimensions are shown in table 9.2. They are standard scores which means that for the city as a whole the average on each dimension has a value of zero and the standard deviation is equal to one. For socio-economic status, positive scores indicate below-average status and negative scores, above-average; the opposite applies for the other two dimensions, positive scores showing above-average status.

The positive socio-economic values therefore show that for all three census years the status of the inner city as a whole was below the city-wide average. Between 1961 and 1971 there was a statistically significant decline in status. The subsequent ten years show a marginal but not significant increase. These results are consistent with expected socio-economic trends prior and subsequent to revitalization in the inner city. The average scores, however, to some extent conceal more substantial changes occurring in specific tracts. For example, a marked decline in status occurred between 1961 and 1971 in tract 37 bounded by King Street,

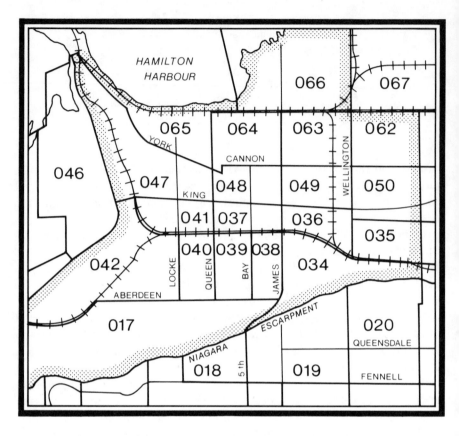

Figure 9.10 Census tracts in central Hamilton, 1981

James Street, the TH&B Railway, and Queen Street, and this decline continued between 1971 and 1981. In contrast, marked increases occurred in 1971-81 in tract 48 bounded by Queen, Cannon, James, and King streets, and in tract 34 between James and Wentworth streets, immediately below the escarpment. These are areas which have experienced new apartment and condominium developments, and, in the first case, the revitalization associated with Hess Village has also been an important factor.

The average scores in the family-status dimension are all negative indicating levels consistently below the city-wide average. This finding is in line with the expectation that the inner city has a relatively high proportion of single- or two-person households and a correspondingly low proportion of households with children. There is some evidence of a return of family households to the inner city based on the higher average score for 1981. No individual census tracts show major decreases or increases in family status over the twenty-year period. Tracts with scores above the city-wide average are generally to the north and east of the CBD, corresponding to the areas which have traditionally attracted immigrant families (e.g., Italian, Portuguese, and Southeast Asian). The inner city also contains tracts which have the lowest family status of any in the city. These are the tracts bounded by King Street, James Street, Queen Street, and the TH&B Railway (37) and by James Street, Aberdeen Street, Bay Street, and the TH&B tracks (38).

Ethnic status in the inner city has generally been above the city average throughout the period (table 9.2). The decline in ethnic concentration from 1961 to 1971 is close to being statistically significant as is the subsequent increase from 1971 to 1981. The latter change reflects the arrival of Southeast Asian immigrants, including the Vietnamese and Cambodian refugees for whom special housing provision has been made in inner-city apartment buildings (e.g., Queen Street North). Tracts with particularly high concentrations of ethnic groups in 1961 were the four north of the CBD bounded by the harbour; the CN railway tracks, Wellington, King, James, and Cannon streets; and York Boulevard (49, 63, 64, and 65). In each case, there was a reduced concentration by 1971 and this was especially true in tract 49, immediately adjacent to the CBD. Tracts showing the largest increases in ethnic status between 1971 and 1981 were three of these same tracts (49, 64, and 65) plus the area bounded by the CN tracks, and Wentworth, Cannon, and Wellington streets (62).

These inner-city tracts emerge as an area of below-average economic and family status and above-average ethnic status. This profile is consistent with that reported in similar studies for other North American cities. This general consistency over time does, however, obscure major changes which have occurred in specific inner-city areas. These changes are most evident for economic and ethnic status and in the former case are indicative of the effects of general trends such as suburbanization and subsequent inner-city revitalization.

CONCLUSION

The recent publication of the 1981 census data provides an opportunity to examine the current social structure of the city and to monitor changes which have occurred in the last twenty years. From a geographical perspective, we are particularly interested in determining the effects of such general urban trends as suburbanization, inner-city decline and renewal, and ethnic dispersion.

Analysis of the 1961, 1971, and 1981 census data for Hamilton yielded scores for each tract on each of three dimensions of social structure: economic, family, and ethnic status. The mapping of the scores for economic status showed a marked eastward and southward extension of higher-status areas previously confined primarily to the west end. New single-family housing construction above the escarpment was the major factor leading to this change. The typical zonal pattern for family status was confirmed for Hamilton, although there was evidence of the pattern weakening over time in part as a result of increases in status for some inner-city areas. A trend towards a greater dispersion of ethnic groups was evident. The traditional area of concentration in the inner-city persisted but other areas of relatively high concentration emerged eleswhere, most notably the eastern margins of the city, bordering Stoney Creek.

More detailed analysis of social characteristics in the inner city showed that the area has been of below-average economic and family status and of above-average ethnic status throughout the study period. There was, however, evidence of a decline and subsequent increase in economic status consistent with the general trends of inner-city decline and renewal. A similar sequence of a fall (1961-71) followed by a rise (1971-81) was shown for ethnic status, the latter reflecting the arrival of new immigrants, including Asian refugees in the last decade. Beyond these general fluctuations in the social composition of the inner-city, more marked changes were observed for specific inner-city tracts. These data almost certainly underestimate the magnitude of localized changes because the census tract covers a sufficiently large area that changes restricted to a particular street or block are diluted in the calculation of tract averages. This artifact of the data explains, for example, why the upgrading of certain blocks within the tracts to the south of the CBD does not show clearly in the results.

From these results, we see to some extent how the trends described in the previous essay are exhibited in the social landscape of Hamilton over the study period, 1961-81. Suburbanization is clearly shown by the growth in population on the southern and eastern margins of the city. Moreover, the social characteristics of this population are consistent with typical suburbia: namely, somewhat above-average economic status and family status. Inner-city decline followed by renewal are shown by the drop and subsequent rise in economic status of inner-city Hamilton as a whole and by the more marked changes which have occurred in specific tracts. Finally, the trend towards ethnic dispersion is supported by the emergence of several areas of ethnic concentration in addition to the traditional core area in the inner city.

REFERENCES

Herbert, D. 1972. *Urban Geography: A Social Perspective*. Praeger, New York
Ley, D. 1983. *A Social Geography of the City*. Harper and Row, New York
Sherky, E., and Bell, W. 1955. *Social Area Analysis*. Stanford University Press, Stanford
Timms, D. 1971. *The Urban Mosaic*. Cambridge University Press, Cambridge

CHAPTER 10

The move from county to region

A.F. BURGHARDT

Nineteen years have passed since the beginnings of the process which transformed our system of local government. Hamilton-Wentworth and its three neighbouring regions now appear to be permanently in place. The calls for a reversion to the old order, so commonly heard in 1978, are now rarely noted. Although the acrimony is behind us, it is well to examine the process, with all its pain and strife, which led from city and county to region.

Under the old territorial-administrative system, sets of townships were lumped together to form counties. Cities were self-governing units separated out of the counties. A sharp rural/urban differentiation was implied. The scarcity of convenient local transportation kept the cities compact and maintained the separation of the city from the countryside. The many villages and general stores satisfied most of the basic economic and social needs of the farmers. City limits marked the edges of the areas of urban development. When the cities expanded, their limits were adjusted outward to include the newly urbanized areas.

The township-county-city system had been largely borrowed from the United States by the reform government of Baldwin-Lafontaine in 1849. However, even though it resembled the American model, it remained within the British structure of power, which held that municipalities were administrative subdivisions of the provincial territory and, as such, subject to supervision and alteration by the higher authority.

The old system became subject to severe strains after 1945 because of the universal acceptance of the automobile and the extension of highways to meet its needs. During the post-war years family living was the norm, and the four-child family growing up amid the quiet greenery of suburbia was considered to be the ideal life. Further, there was a universal acceptance that urban-level services should be the norm wherever possible. For a decade and a half the provincial government attempted to fit the new developments within the old model, and adjusted the city limits to match the areal dimensions of explosive urban-suburban growth. By 1960, the government and its planners had concluded that this was a hopeless policy, in that it led to the mutilation of the municipalities around the cities, and in that there was no hope of keeping up with the range of the commuters.

The extension of urban uses beyond the legal city limits had led to conflicts between neighbouring municipalities. The formerly rural townships found themselves under severe financial pressures because of the need to supply services to an expanding young population, while yet lacking the assessment needed to meet the costs. Meanwhile the central city found itself expected to pay for the higher-level services which would be used, but not paid for, by people living beyond the city limits. The cities were also left with most of the welfare and redevelopment costs. Planning had become a necessity, and it became apparent that the policies of the central city and of the surrounding county had to be harmonized, despite continual competition by all units for industrial and commercial assessment.

By the 1960s then, the Government of Ontario had come to the conclusion that a revamping of the local-governmental system was imperative in Southern Ontario. The British tradition, spelled out in the British North America Act, granted the government the power to do so, and the uninterrupted rule of the Progressive Conservative party allowed it the luxury of formulating long-range plans, with little fear of electoral defeat. A series of initiatory studies culminated in the Smith Report on Taxation (1967), which examined the financial dimensions of local government, and suggested a set of new administrative units to be called 'regions.' Meanwhile similar reorganizations were being carried through in the United Kingdom. The Redcliffe-Maud (1969) and Wheatley (1969) reports for England and Wales, and Scotland, respectively, had a strong influence on plans and procedures in Ontario. However, whereas the English and Scottish proposals treated their entire countries as units subject to new systems of subdivision, the Ontario approach was ad hoc, one local area at a time, with no overall plan for the province.

One idea, that of the city-region, came to dominate planning on both sides of the Atlantic. Formerly the city and its surroundings had been seen as distinctly separate areas, with differing economies and lifestyles. Now an organic unity was perceived, joining the city and its 'hinterland,' or 'service area.' The principal change to be made was the reintegration of the city into its surrounding county or counties. Populations which were seen to function together should be administered together; the functional unit should become the governmental unit. The Government of Ontario accepted this prevailing wisdom and assumed that newer, larger units designed for administrative efficiency would bring about lower costs, and would also diminish the staggering number of requests for aid being brought to the Ministry of Municipal Affairs by ailing local units. Planning would be unified and rationalized.

The provincial government did not seize the opportunity to redraw the entire area, or even the entire Golden Horseshoe, as had been done in Scotland. Instead, the government opted for the expensive procedure of commissioning a separate study of each major component of urbanized Southern Ontario. Not only was this approach very costly, but it also meant that each commission had to have its areal terms of reference stated; inevitably these were based on the units already in existence – the counties. It was not surprising, therefore, that some politicians came to view regional government as little more than an updating of county government.

THE STEELE COMMISSION

A three-man commission, headed by the Toronto lawyer Donald Steele, met through the winter of 1968-9, searching for data and eliciting local opinion. Over one hundred briefs were submitted to the Hamilton-Wentworth- Burlington Area Commission. Although financial matters were given great emphasis, it became apparent that there were three principal questions to be answered: 1 / where should the outer limits be placed and what will be included? 2 / what shall the internal structure be? and 3 / how shall internal political power be distributed?

It was assumed by everyone that all the municipalities of Wentworth County would be included, even if they dreaded the prospect. The major issue was the future position of Burlington. The brief of the City of Hamilton (1968) called for the inclusion not only of Burlington as a part Hamilton urban area, but also the lower course and mouth of Bronte Creek in Oakville on the grounds that the entire creek watershed should be in one region. (Cynics were quick to point out that the area between Burlington and the creek mouth contained the assessment-rich Bronte refineries.) The detailed County of Wentworth (1969) brief also called for the inclusion of Burlington. For the rural and suburban municipalities the addition of Burlington was held to be absolutely essential, because it was the only large unit which could hope to counter the overwhelming population dominance of the central city. However, Burlington, with its vociferous mayor as spokesman, claimed that it would have nothing to do with any new region centred on Hamilton. It was in Halton County, in another school system, and was the principal contributor to the maintenance of that county.

In a somewhat analogous position on the east was Grimsby, which was within Lincoln County but formed the eastern end of the Hamilton industrial belt. Perhaps because it was considerably smaller and had no hopes of becoming a great city, as Burlington seemed to have, Grimsby did not fight against inclusion in Hamilton-Wentworth. Elsewhere, the county brief was content with the county limits. The city brief, however, dipped southward across the Grand River to include Caledonia. Perhaps as a compensation for its suggested expansionism in Bronte and Caledonia, the city offered to give away the northwestern corner of Beverly Township to Waterloo. Curiously, this offer, which was obviously made without consulting the people of Beverly, was the only deviation from the county limits accepted by the provincial government.

The fundamental question concerning internal structure was whether it was a one-tier system or a two-tier system. A single tier meant one level of administration, which in turn meant a unified region, a greatly enlarged city. Two tiers meant a federal structure with two levels of administration. Each local municipality would have its own council, but above these would be a regional council, composed of members from each of the participant units.

From the start, the city urged the adoption of the one-tier system on the grounds of efficiency and economy. This system would, of course, have given the city total dominance of the region and would have meant either the eradication of the outer municipalities or their diminution to the status of wards. Consequently, the 'outer units' (a term which will be used to describe all the municipalities of Wentworth County, except Hamilton) pushed strongly for the adoption of the two-tier system.

It is difficult to overestimate the importance of this matter to the outer units. Their very existence was at stake; it was literally a life-and-death issue, and as such elicited powerful emotions. Since the city continued to press for a one-tier system, even after the region was already functioning, an animosity of profound proportions developed between city and suburb. Distrust of the central city is endemic in North America in any case; Hamilton's continuous campaign to absorb its neighbours exacerbated this distrust.

The two-tier system was the system commonly in use in both the United Kingdom and elsewhere in Southern Ontario. Metro Toronto was a successful pilot case of this structure. While efficiency seemed to demand the establishment of regional councils, local democracy seemed to demand a lower level of participation as well.

Other vital questions concerned the number and size of the component municipalities, if a two-tier system were to be selected. Although never clearly stated, the commission did operate within threshold assumptions which had been drawn from the Smith Report (1967). At least 10,000 inhabitants appears to have been the lower limit; however, few of the outer units reached that total. Were they to be merged? If so, how? And under which names? Each municipality possessed a powerful will to live; each could call forth almost two centuries of traditions and local associations. Even if merging were to be decreed, the local name should not be erased from the map.

There was also the belief that there was a fundamental dichotomy between urban and rural-farming life. The city-region concept might postulate a unity, binding the two together, but the lifestyles, views of the land, and approaches to taxation were held to be radically different. However, although this distinction was strongly held by the outer units, it was either ignored or felt to be unimportant by the city. Both the County Brief (1969) and the ensuing Steele Report (1969) allowed for a separation of the region into a smaller urban core and a larger (areally) rural periphery.

The question of how power was to be distributed came up against the overwhelming numerical majority enjoyed by the city. Without Burlington and Grimsby, the city would hold three-quarters of the population; even with them the city would include approximately two-thirds. The outer units believed, however, that the city could not be allowed a majority of the members on the regional council. It was assumed that the city members would vote as a block on key issues, and thus threaten the continued existence and viability of the outer units. The County Brief (1969), which called for the inclusion of Burlington and Grimsby, called for an equality of representation between the city and the outer units.

The central issue was thus a clash between the principle of representation by population and the principle of equal representation for major components of the region. Strict representation by numbers would adhere to the concept of one-person–one-vote, but also allow for 'the tyranny of the majority.' The farmers would be in the ironic situation of occupying most of the land, but of having virtually no voice in the regional council which would govern the use of that land. Rural people believed absolutely that urbanites, and their planners, would have no understanding of or sympathy for rural concerns. The Steele Commission (1969) could not bring itself to accept an equality so much at variance with numbers, but did allow the outer units a higher number of regional councillors than numbers would dictate.

The Steele Commission presented its report in November 1969. It called for a two-tier structure, including both Burlington and Grimsby. A host of recommendations spelled out the distribution of powers and responsibilities, and the control of finances.

The report met with the immediate strong opposition of the elected officials of Burlington. By coincidence this opposition occurred just prior to the civic elections. A referendum on regional government was hastily thrown together and placed on the ballot in Burlington. This was highly irregular in that the provincial government had explicitly stated that referenda were not to be allowed, and, in fact, a referendum had been banned at the Lakehead. The question on the ballot mentioned no alternatives; it merely asked whether or not the voters wished to become a part of Hamilton-Wentworth.

Fewer than half of the eligible electors voted on this issue, but, as could be expected, the vote was overwhelmingly against the proposal. Somewhat similar votes held elsewhere have had similar results; in truth, no North American suburb will vote to join the central city. Suburbs such as Leaside (Toronto), Meritton (St Catharines), and Streetsville (Mississauga) have accepted absorption only because it was forced upon them. Nevertheless, the vote had been held and supplied statistical witness of local wishes. This evidence, plus the presence of a Burlington MPP within the provincial cabinet, supplied the ingredients for a successful Burlington opposition to inclusion in the Hamilton-Wentworth region.

Except for continuing disputes between Hamilton and the outer municipalities, nothing further happened for more than three years. The provincial government seemed to let this troublesome area languish while it went forward with plans for the neighbouring areas. First to be organized was the Regional Municipality of Niagara, formed out of what were previously the counties of Lincoln and Welland. Since it was a part of Lincoln, Grimsby was automatically included. When Hamilton representatives complained that this ran counter to Steele's findings and recommendations, the government replied, with impressive duplicity, that nothing was final; Grimsby could be transferred later. The establishment of Waterloo was also underway, and the government planners seemed intent on adding the northwestern corner of Wentworth to Waterloo. Instead of adding territory, Wentworth seemed to be in the position of being continually cut back.

By 1972, it was becoming increasingly apparent that Burlington was not to be included within a region centred on Hamilton. This realization created a sense of near-panic among the representatives of the outer units. Desperate to find a counterweight to the dominance of Hamilton they sought to include Brant County and the City of Brantford in the new region. Proposals to that effect were formulated and forwarded to Queen's Park, but were probably never taken seriously.

Finally, in January 1973, the provincial government was ready to announce its proposals for the shoreline counties west of Toronto. The elected officials of Peel, Halton, and Wentworth counties and their cities were invited to a meeting held at Mohawk College in Hamilton. In step-wise fashion, the deputy minister of municipal affairs moved westward from the borders of Metro Toronto. There was to be a two-tier Region of Peel, another of Halton, but two possibilities were set forth for Hamilton-Wentworth.

It was apparent that Municipal Affairs had viewed the organizational

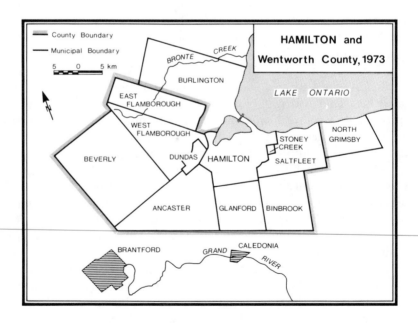

Figure 10.1

problems as seen outward from Toronto; Hamilton was not perceived as a major central city whose hinterland was to be defined, but only as a part of the Toronto-centred region (see Government of Ontario, 1970). Since Metro Toronto was not to be enlarged, the rapidly growing area just to its west would have to be established as a region. Was this to be Peel County alone or Peel plus a part of Halton? Peel was certainly large enough in population and in potential growth to be a separate region. Making the new region conterminous with the former county would clearly cause the least disruption. Furthermore, Peel elected officials had rebuffed previous attempts by Halton to join Peel, and there was the fortunate coincidence that the premier of Ontario was from Peel.

Once Peel had been established as a region, Halton had to be kept intact. Adding Burlington to Wentworth would have left a rump county, or region, oddly shaped and deprived of a major portion of its population and assessment base. In effect, Burlington could be joined to Hamilton-Wentworth only if the remainder of Halton were joined to some other region. Here, too, then, the Steele recommendations were ignored.

Faced with persistent internal squabbling, the provincial government cleverly brought forward two proposals for Hamilton-Wentworth, only one of which could be deemed acceptable. By this device, the government in effect forced the disputing antagonists to accept the plan which was being introduced elsewhere, while still making its acceptance look like the free choice of the representatives involved.

The first option called for a one-tier system but with a major amputation of territory. All the area north of Highway 5 was to be lost: that west of Highway 6 to Waterloo, that east of the highway to Halton. The scale of the loss was stunning; such local features as the Plainsman Restaurant and Flamboro Downs racetrack would have gone to Waterloo. The north-western corner of Clappison's Corners, only 8 kilometres (5 miles) from the centre of Hamilton, would have been administered from Kitchener, 56 km (34 mi) away!

The second proposal was for the two-tier system, which, with minor changes became the basis of the new Region of Hamilton-Wentworth. Although there was no counterweight to the dominance of the central city, the spokesmen of the outer municipalities accepted this alternative. Yet, there were city representatives who spoke favourably of the first option, simply because it promised a one-tier system.

A number of matters had still to be worked out. Despite the refusal of the first option, Municipal Affairs continued to be intent on severing the northwestern corner of Beverly Township and giving it to Waterloo, because of its proximity to Galt-Cambridge. The initial proposals would have involved the loss of the African Game Farm and Wentworth Pioneer Village. Local protests led to a reduction of the area to be lost, to around thirty square kilometres (nineteen square miles). Beverly lost even its identity, because of its low population. Despite its large area, it was merged with the two Flamboroughs, and witnessed the disappearance of its cherished name from the map. Further, Flamborough's new town hall was erected well to the east, near the centre of town population, but many kilometres distant from the western reaches of Beverly.

The governmental proposal called for a union of Dundas and Ancaster. As seen from the planners' desks in Toronto, the two historic towns must have seemed to be no more than neighbouring suburbs which could easily be merged. Local efforts persuaded the government that the two were indeed distinct locationally, historically, and economically. Thus, Dundas became the only municipality of small area, and hence the only one to maintain its small-town character despite the regional bias towards bigness.

Nomenclature, the naming of the new municipalities, could also raise strong feelings. Stoney Creek and Saltfleet were to be merged; what should the new unit be called? Saltfleet had virtually all the area, most of the population, and most of the assessment, but the name of Stoney Creek had been sanctified by history (the War of 1812), and could not be allowed to disappear from the map. In the case of the two units to the south, Glanford and Binbrook, a simple compromise was arranged: one syllable from each name. 'Glanbrook' was clearly preferable to 'Binford,' so Glanbrook it became.

The new Region of Hamilton-Wentworth, which came into existence on 1 January 1974, was composed of six municipalities, each of which continued to have its local town (or city) council (Government of Ontario, 1973). Hamilton had seventeen members, while each of the five outer units, despite great differences in population, had two members each on the regional council. Certain functions, such as water and police were unified on the regional level, but most, including planning, fire protection, and recreation, were both regional and local.

AFTERMATH

It quickly became apparent that the enforced unification of the city and most of the former county did not guarantee a harmonious relationship, despite the laudable efforts of the regional chairman to achieve one. The outer municipalities remained distrustful and fearful of the city; they were

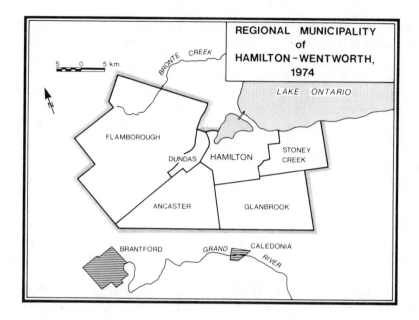

Figure 10.2

desperate to maintain their identities. Their loyalties were to their own units and not to some newly imposed entity. The idea of working together for the common good of the region stood little chance against the fundamental purpose of protecting the viability of the individual units. Some city representatives exacerbated these tendencies by continuing to advocate a move towards one tier. The city-county animosity was heightened by a continual clash of personalities. One of the two regional representatives from each of the outer municipalities was the mayor of that municipality. These men and women were strong, outspoken individuals who felt a unique responsibility to protect the interests of their home units. In these first years, the mayor of Hamilton was likewise strong and outspoken, with an ability to antagonize the non-city representatives. Confrontation rather than co-operation seemed to be the rule. Much of the rancour concerned financing and planning.

Regionalization brought with it a startling increase in costs which was unexpected because of the prevailing assumption that bigness meant greater efficiency and hence savings in administrative costs. In fact, bigness led to bigger structures: more office space, more staff, and higher-salaried professional managers. This sudden rise in costs was not unique to Hamilton-Wentworth. Similar increases occurred in all the neighbouring regions, and had occurred in the school systems when they were unified a decade previously. Bigger meant more expensive because bureaucratic growth always exceeded any gains realized from local cut-backs..

Each faction tried to use the increase in costs as a weapon. The city maintained that the expenses were due to the duplication of services on the two levels and that a move to a one-tier system was needed. The outer units claimed that the rise in costs was the fault of the new system and could be solved only by a reversion to the old system, which was, in fact, still functioning amicably in neighbouring Brantford.

The greatest animosity resulted from the questions as to whether the entire region or only the city would pay for those central-city cultural and social functions which served the entire region. Feeling became so strong that the representatives of the outer municipalities walked out of regional council meetings in order to stop proceedings. Business could be conducted only if representatives of three of the six municipalities were present. Since on these issues the city representatives did indeed vote as a block, the representatives of the other units felt that their only recourse was to walk out and thus prevent council from reaching a decision.

A regional plan was to be completed and submitted to the Department of Municipal Affairs within three years. This deadline could not be met because of the need for the planning office to accommodate the wishes of the six municipalities. Regional planning and a doing away with destructive competition for assessment had been among the principal reasons for the administrative reorganization. However, the internal financial structure did not do away with the competition for assessment. Each municipality contributed to the regional costs a share of its tax income. In effect, the higher its tax base, the more it paid. But if local assessment were increased, the municipality would still gain about half of the increase (much as an increase in personal salary will mean an increase in income tax but nevertheless still yield more funds for the individual). Thus, every one of the municipalities demanded an allowance for the creation of an industrial park within its limits.

It should be pointed out, though, that internal squabbling became the rule in almost every region of Southern Ontario. Attempts at secession became commonplace. In nearby Niagara, St Catharines repeatedly tried to leave the region, as did Cambridge in Waterloo, and Milton in Halton.

Inevitably, regionalization became an issue in the provincial election of 1975. The Liberal party called for the dismantling of the regional governments and made significant electoral gains. For the first time in over three decades the Progressive Conservative party lost its majority in the provincial legislature. It announced then that no more new regions would be established unless the local people specifically requested one. Needless to say, no such requests have come forth. This meant, in effect, that such urban centres as London, Windsor, Brantford, Kingston, Cornwall, Peterborough, and Barrie, and their surroundings would continue to function within the old township-county-city system. Local politicians could hardly be blamed for concluding that what was good enough for London and Middlesex County should surely be good enough for Hamilton and Wentworth County.

There were repeated calls for a review (as there were in other regions, too). Hamilton wished to move towards a one-tier system, whereas the outer units wished to return to the old system, which was increasingly being looked upon as a kind of Golden Age. After several years of refusal, the provincial government established the Stewart Commission in 1978 to review the functioning of Hamilton-Wentworth. Once again there was the call for briefs, the reception of delegations, and the analysis of available data.

The Stewart Report (1978) came as a profound shock to the representatives of the smaller municipalities. Shouts of 'assassin' were heard in the hall, as Mr Stewart read out his recommendations, calling for a change to a one-tier system. The distrust of the central city was to be exorcised by changing the name of the new super-city from Hamilton to Wentworth. A new ward system was to be introduced but not on a per-capita basis. The new wards were to be placed within the existing municipal boundaries and the number assigned to each seemed to depend as much on area as on population. A few of the wards were designated as 'rural.' Despite the disparity in population among wards, Hamilton was pleased with the report because the city's desired one-tier structure had been recommended.

The outer five units were enraged. Not only had they lost their hope of a return to the golden past, but they had seen the recommendations go to the opposite extreme. An emotional campaign was launched, most strongly in Dundas, against the implementation of the report's recommendation. In the end, the provincial government agreed to shelve the report and maintain the system already functioning in the region.

In retrospect, one can see that the governmental actions in 1978 were a repetition of those of 1973. Once again the region had been presented with two 'options,' one of which was totally unacceptable to the outer units. Once again the rebellious factions were forced to accept the two-tiered region as the lesser of two evils. They found themselves thrust into the position of clamouring vociferously for the very system which they had previously denounced.

To be fair, one should not impute any duplicity to Mr Stewart and his colleagues. The commission's terms of reference did not allow for any dismantling of the region or any alteration in its outer limits. Its mandate was to seek improvement. Since he could not go back, and the present was not functioning smoothly, he had to move 'forward' to a unified structure, in the hope that the abolition of 'duplications' would yield fiscal savings, and would force all the units to work together. The region has, in fact, worked together more smoothly since 1978. The City of Hamilton elected a new mayor and co-operation rather than confrontation began to characterize the meetings of the regional council.

AFTERTHOUGHTS

With the clarity of hindsight, one may draw a few conclusions from this history of change. It is quite clear that local loyalties remain distinctly local. Even after a decade, it seems safe to say that identification remains with Hamilton, Dundas, Ancaster, Stoney Creek, Flamborough, and Glanbrook, and probably with Beverly, rather than with the region. Regional councillors, all of whom are elected from within their municipalities, cannot help but place top priority on the interests of their local units.

It is also clear that not the government, or its planners, or the city officials, or the press had any appreciation of the depths of local territorial feeling. Territoriality – the close, emotional, almost instinctive ties between people and certain tracts of land – can be a powerful force, especially when it feels itself to be threatened. There seemed to be an assumption that the opposition to reorganization was either a neanderthal reaction to change or the selfish opposition of local politicians. It was apparent, especially in 1978, that a deeply emotional identification with the old units included a large segment of the population. In some cases non-elected inhabitants were more vociferous than their representatives, because the latter were cowed by the higher levels of government.

A sense of local tradition is very strong in smaller communities. It may have been a misfortune for the government that it introduced this wrenching transformation just as the preservationist movement was gaining strength. Whereas for the central planners and administrators, economics was considered to be more important than history, for articulate local people history was more important than economics. The clash between these viewpoints gave rise to deep emotions and the feeling that local democracy had been flattened by a bureaucratic steam-roller. Names were perceived as being of great importance because they were the concrete expressions of local identity. Beverly, the largest township in area, lost its name, its central facilities, and even a piece of its territory.

Finally, one can't help but conclude that Hamilton-Wentworth was treated more roughly by Queen's Park than were its neighbours, some of which were just as recalcitrant and full of internal conflict. Ottawa received all its suburban areas within Ontario (Hull, as part of another province, could not be included in the Ottawa-Carleton Region). St Catharines, Kitchener, and Sudbury received all their suburban areas, whereas Hamilton-Wentworth was kept separate from Burlington. This has proved to be a serious omission because much of the recent expansion in both industry and housing in the Hamilton area has been along the Queen Elizabeth Way in Burlington; further, many of the area's social leaders live in Burlington and, hence, outside the region. Among the regions, only Wentworth and Halton lost territory. (Oakville, in Halton, lost a thin strip of territory to Peel to allow for the expansion of the Sheridan Park industrial-research area.) As described above, one of the options of 1973 would have awarded almost one-third of Wentworth to Kitchener-Waterloo and moved the Waterloo-Wentworth boundary to within 3 km (1.8 mi) of the Hamilton city limits. One is tempted to conclude that there may have been a planners' bias in favour of Kitchener over Hamilton.

REFERENCES

City of Hamilton. 1968. *Submission* [to the Hamilton-Burlington-Wentworth Local Government Review Commission], Hamilton

County of Wentworth. 1969. *Brief* [to the Hamilton-Burlington-Wentworth Local Government Review Commission], Hamilton

Government of Ontario. 1970. *Design for Development: The Toronto-Centred Region*. Queen's Printer, Toronto

– 1973. An Act to establish the Regional Municipality of Hamilton-Wentworth, 1973

Redcliffe-Maud, the Rt. Hon. Lord. 1969. *Report*. Royal Commission on Local Government in England, 1966-69 (CMND 4040). HMSO, London

Smith, L.J. 1967. *Report*. Ontario Commission on Taxation, Toronto

Steele, D.R. 1969. *Report and Recommendations by D.R. Steele (Hamilton-Burlington-Wentworth Local Government Review)*. Ontario Department of Municipal Affairs, Toronto

Stewart, Henry E. 1978. *Report of the Hamilton-Wentworth Review Commission*. Queen's Printer, Toronto

Wheatley, the Rt. Hon. Lord. 1969. *Report*. Royal Commission on Local Government in Scotland, 1966-69 (CMND 4150). HMSO, Edinburgh

PART 3

HOW
HAMILTON
WORKS

Downtown Hamilton: the reconstructed Gore Park
(photo by R. Bignell)

Downtown Hamilton: apartment buildings
(photo by R. Bignell)

Left, top: Ontario Hydro substation
Left, bottom: Stelco and Dofasco
(photos by R. Bignell)

Farming within the city limits
(photo by R. Bignell)

Caring for the elderly
(photo by R. Bignell)

CHAPTER 11

Farming in an urban age

L.G. REEDS

Soil and climatic capability and technological and socio-economic factors have all affected the evolution of agricultural land use in the Hamilton region and in the Niagara fruit belt. The number of farms in the Hamilton-Wentworth region has declined by 52 per cent, from 3,265 in 1941 to 1,553 in 1981, and the area of occupied farmland by 36 per cent, from 100,190 ha (247,570 ac) in 1941 to 64,150 ha (158,515 ac) in 1981 (Census of Agriculture). Concurrent changes include increases in the size of farms, in part-time farming, tenancy, land values, mechanization, and in capitalization. These developments relate to adaptations to the physical environment, adjustments necessitated by the expansion of Hamilton, and changing economic conditions and technology. The great advancement of productivity which occurred in this time period was accompanied by vast increases in capitalization and in production costs. The selling price for produce has not kept pace with input costs in recent years and has brought about farm bankruptcies.

The first topic to be examined in this essay is a basic one, that of the relationship between agriculture and the physical environment. As outlined in chapter 4, the region has a great variety of soil types that differ in slope, drainage, texture, and agricultural capability. These variations relate to the differences in the glacial materials on which the soils have developed during the past 12,000 years since the ice sheets covered the area.

The most significant climatic difference is associated with that portion of the Niagara escarpment that is known locally as Hamilton Mountain. The area between the escarpment and Lake Ontario has a longer frost-free season than that to the south, a condition that partly accounts for the concentration of tender tree-fruit production in the lowlands bordering the lake. The climatic differences within the area above the escarpment are of less significance and have not been a major factor in influencing variations in agricultural land use.

For the most part, the climate (with its 3,800 growing degree days, based on the Fahrenheit scale – that is, the sum of the number of degrees the daily mean is above 42°F, or 5.5°C – and adequate precipitation) is ideal for the growth of a wide variety of general and special field crops and ranks as one of the most favourable in Canada. The seasonal changes in temperature are shown in figures 3.4 and 3.5 in chapter 3.

AGRICULTURE AND SOILS

The soils of the region have been classified on the basis of their inherent capability into seven classes. Class 1 soils have no important limitations for agricultural use. Limitations increase from Class 1 to Class 7, the last class being unsuitable for agriculture (Ontario Ministry of Agriculture and Food, 1965). The prime farm land includes the soils in classes 1, 2, and 3. As much as possible of this land should be conserved permanently for food production. Class 4 soils are not suitable for sustained production of cultivated field crops and are suitable only for a few crops as yields are lower and the costs of production higher. Classes 5 and 6 have many limitations and are best used for pasture. One should keep in mind that this classification is based entirely on inherent characteristics and does not include socio-economic criteria that may affect agricultural production. Such criteria as distance to an urban centre, size of farm, land ownership, cultural patterns, and the skills and resources of individual operators are not included.

The distribution of the soil capability classes of the Hamilton region is shown in a generalized way in figure 11.1. Generally speaking, the best soils are the well-drained heavier-textured soils that occur on the gentler slopes. These soils have good water-holding capacity, are naturally well supplied with plant nutrients, are easily maintained in good tilth and fertility, and are adaptable for a wide range of general field crops (Ontario Ministry of Agriculture and Food, 1965).

Four of the more extensive prime farmland soils include those of the Brantford, Beverly, Binbrook, and Grimsby series. The Brantford and Beverly silt loams are inherently well suited for the production of many crops including forage types, spring and autumn grains, and corn and canning crops. They occur in the southwestern parts of the region. Specialization in dairy farming in the township of Glanbrook has led to a concentration of forage crops and grain on the Binbrook soil types. The

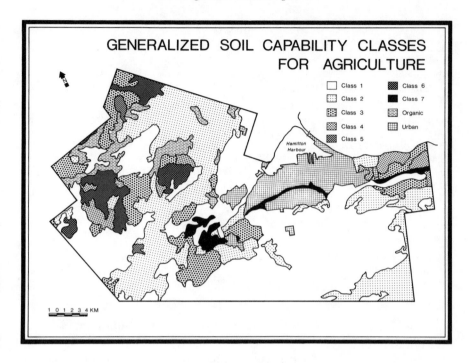

GENERALIZED SOIL CAPABILITY CLASSES
FOR AGRICULTURE

Class 1   Class 6
Class 2   Class 7
Class 3   Organic
Class 4   Urban
Class 5

Hamilton
Harbour

1 0 1 2 3 4 KM

Figure 11.1 Soil capability in Hamilton-Wentworth

specialized cash crops – grain corn, sweet corn, canning peas, and to-matoes – are also important. Artificial drainage has improved productivity in these three series, which have developed on heavy-textured lacustrine or till materials. Grimsby soils, in contrast, occur on well-drained sandy loams. Because of their texture, they do not hold moisture as well and are prone to droughtiness. While not quite as productive as the heavier-textured soils for forage crops and grain, they are good for sweet corn, tomatoes, strawberries, and, in the Niagara region, for peaches and sweet cherries.

Limitations of soils of classes 4 to 6 relate to low inherent fertility, deficiencies in the storage capacity or release of soil moisture to plants, poor structure or low permeability, erosion, topography, overflow, wet-ness, stoniness, and depth of soil to consolidated bedrock. Class 5 soils include the steeply sloping, poorly drained, and shallow types. For the most part, these limitations preclude the production of annual field crops. The soils may be improved and used for perennial forage plants. Class 6 soils do not warrant improvement practices and can only be used as a

TABLE 11.1
Acreages of soil capability for agriculture*

Regional Municipality of Hamilton-Wentworth	Class 1	Class 2	Class 3	Class 4	Class 5	Class 6	Class 7	Organic soils (0)	Totals
Ancaster	16,134	8,908	13,926	–	110	–	4,472	330	43,880
Dundas	1,510	840	1,120	–	–	–	1,940	–	5,410
Flamborough	22,128	30,143	19,566	6,365	15,550	11,567	4,669	10,852	121,110
Glanbrook	22,495	9,880	16,025	3,080	120	80	–	–	51,680
Hamilton	11,197	6,756	4,316	375	751	125	–	–	23,520
Stoney Creek	4,292	7,274	5,697	2,971	1,188	385	2,633	–	24,440
Unmapped	–	–	–	–	–	–	–	–	5,160
TOTALS	77,756	64,071	60,650	12,791	17,719	12,157	13,714	11,182	275,200
PERCENTAGES	28.2	23.3	22.0	4.7	6.4	4.4	5.0	4.0	100.0†

SOURCE: Ontario Ministry of Agriculture and Food, *Report* no. 8
* 1 acre = 0.405 hectare.
† Includes 2.0 per cent unmapped.

natural grazing area. Class 7 soils have such severe limitations that they are not suitable even for pasture. In some cases, they may have a high capability for trees, native fruits, wildlife, and recreation. They include the ravines, marshes, stream courses, rocky areas, quarries, and the escarpment.

Table 11.1 shows the acreages of soil classes in the Regional Municipality of Hamilton-Wentworth. From the standpoint of soil capability, the region has a relatively favourable physical base, with 73.5 per cent of its area classified as prime farmland. It is unfortunate that a large proportion of this most productive land lies in the vicinity of Hamilton where urban growth has and will continue to create conflicts in land use (figure 11.2). Mount Hope airport is located on prime farmland. Urban sprawl that has engulfed large areas of farmland in the northern part of Stoney Creek is a less serious development since much of this land is only marginal for tender-fruit production and is not first class for general crops. It should have been provided with services at an earlier date in order to permit industrial, commercial, and residential development.

TRENDS IN AGRICULTURE

Trends in farming in the Hamilton-Wentworth region are similar to those which have been occurring in Southern Ontario during the past forty years.

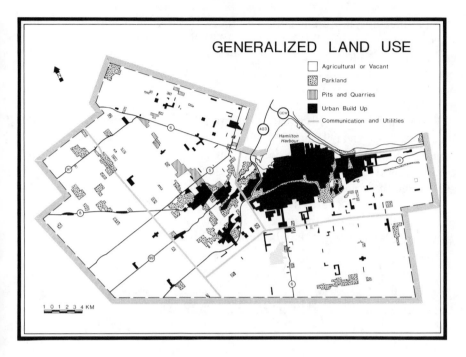

Figure 11.2 Land use in Hamilton-Wentworth

Since 1945, farms have become larger, more mechanized, more specialized, and more highly capitalized. With the adoption of new technology
and the availability of scientific knowledge from the agricultural colleges
and research agencies, the volume of agricultural output has increased in
spite of the declining acreages and number of farms.

In our technological society, the space demands for industry and the
associated housing and services, for transport facilities, and for waste
disposal and recreation have resulted in a drastic decline in the area
occupied by farms. In order to remain economically viable, the agricultural
industry has to concentrate on the best soils in the most climatically
favourable areas. As the soil-capability map (figure 11.1) shows, much of
the land directly to the south of Hamilton is prime farmland, and the
climate rates as one of the most favourable in Canada for a great variety
of crops. Thus, as the city expands southward, prime farmland is lost.

Mixed farming, the dominant type in 1941, has been replaced by specialization in dairying, beef, hogs, poultry, and horticultural and cash
crops. With specialization, farm management has advanced and agricul-

ture has become more efficient. The index of production has risen from 100 in 1961, to 128 in 1971, and 133 in 1975 (Roberts, 1983). For example, grain-corn yields increased 39.5 per cent from 1959 to 1965 and by a similar amount from 1966 to 1970 (Rodd, 1976). However, increases in output per acre and per person cannot be expected to continue at the same rate in the future.

Farms under 4 ha (10 ac) declined from 492 to 231 between 1941 and 1981 and those over 97 ha (240 ac) increased from 31 to 150 during the same period. This change is related in part to the great increases in capitalization and the necessity of enlarging the operation in order to take advantage of economies of scale.

With respect to changes in types of crops and livestock (table 11.2), the most spectacular is the increase in grain corn, a crop that was scarcely grown in 1941 but which has become the leading grain with over 12,000

TABLE 11.2
Changes in acreages of crops* and numbers of livestock
in Hamilton-Wentworth, 1941-81

	1941	1951	1961	1971	1981
CROP					
Wheat	13,659	17,732	9,885	5,785	12,307
Oats	27,216	30,448	32,933	12,465	6,403
Barley	3,123	1,135	806	10,081	7,631
Corn for grain	225	2,062	3,516	17,017	30,607
Rye	298	557	270	220	424
Soybeans	–	169	501	461	3,820
Alfalfa	–	2,638	–	–	17,220
Potatoes	4,191	2,098	2,878	2,032	2,030
Vegetables	–	2,969	3,922	5,768	7,780
Nursery products	–	183	181	481	1,305
Tree fruits	5,911	5,985	5,012	3,762	2,754
Small fruit and grapes	2,857	3,815	3,138	2,237	1,733
LIVESTOCK AND POULTRY					
Total cattle	31,605	29,974	34,343	30,205	23,503
Dairy cattle	16,660	15,751	15,449	7,917	6,273
Beef cattle	517	1,114	–	–	2,590
Hogs	–	–	30,079	48,886	44,879
Sheep	4,821	2,237	3,455	2,302	4,054
Horses	8,383	3,843	1,150	1,372	1,372
Poultry	456,697	660,714	843,452	1,507,627	2,651,428

SOURCE: Statistics Canada, Census of Agriculture
* 1 acre = 0.405 hectare.

ha (30,000 ac) in 1981. The shift from oats to corn is related in part to technical change. Improved varieties, yields, machinery, and the adoption of corn as a cash crop and as an important feed for livestock brought about the change. Other significant changes include the introduction of soybeans and increases in the area of alfalfa, vegetables, and nursery products, including sod for landscaping.

In the case of livestock, the number of dairy cattle has declined by 60 per cent but total milk production has been maintained by increases in output per cow through improvements in the quality of the animal and in feeding. Horses utilized for work purposes have virtually disappeared and are used now mainly for riding or for racing. Statistics show increases in beef cattle, hogs and poultry, changes which are related to the trend from mixed farming to more specialized types of production.

Other important changes relate to ownership patterns, part-time farming, and land values. The total area of owner-operated farms decreased by 51.1 per cent from 84,785 ha (209,509 ac) in 1941 to 41,226 ha (102,365 ac) in 1981, while the amount of rented land increased by 47.5 per cent, from 15,405 ha (38,066 ac) in 1941 to 22,726 ha (56,157 ac) in 1981. The number of part-time farms (defined as those for which the farmer derives 50 per cent or more of his gross revenue from off-farm work), increased from 139 in 1941 to 869 in 1971. Land values vary in accordance with size of farm, quality of land, and proximity to the city, but generally increased sixfold between 1941 and 1981. These changes associated with part-time farming, type of tenure and value of land reflect changes in the structure of farming that have been occurring in agriculture since 1941 with the trend to mechanization, increasing capital costs of operation, and land speculation.

The rural-urban land conversion merits closer examination. Annexations by the city of Hamilton southward and eastward have enveloped all the farmland in the former township of Barton and large areas in the former township of Saltfleet. As the existing land-use map (figure 11.2) shows, urban sprawl has reached its maximum extent in the northern part of the town of Stoney Creek. Public and private open spaces occupy relatively large areas in the vicinity of Dundas and Ancaster and in the township of Flamborough. Pits and quarries are a conspicuous feature of the landscape in the limestone plain and in the gravelly moraines of Flamborough. The scattered roadside residential development throughout the municipality is a mute testimony to the haphazard development that occurred before planning was introduced in the 1950s.

LAND USE–DISTANCE RELATIONSHIPS

The problem of urban expansion in the rural-urban fringe has been extensively documented (Beesley and Cocklin, 1982). Many researchers have described the extent of non-farm development but few have analysed the causal factors (Reeds, 1969). Much of the urban sprawl in the Hamilton-Wentworth and Niagara regions consists of single-family residences and scattered subdivisions along roads and highways.

The large price differential between lots in the city and in rural areas entices some buyers to choose a rural location in order to achieve the apparent savings. However, in moving farther from the place of employment, the cost of commuting is increased proportionally. As the cost of commuting increases with distance, the amount actually saved as a result of lower land costs is reduced. At some point beyond the urban area, the increased costs will equal the savings. Assuming rational economic behaviour on the part of buyers, this point should be the limit of residential development beyond the city. Farmers living near cities seek urban employment and benefit from the lower costs of rural living. This off-farm employment is an increasing trend throughout Southern Ontario but is especially prevalent in the vicinity of Hamilton and in the Niagara Peninsula where distances to several cities are not great.

Thus land use in the Hamilton-Wentworth and Niagara regions is affected in two ways. Some land will be demanded by residential land buyers and its use is converted directly from agricultural to residential uses. Other land is used by farmers (who commute to the city) for agriculture on a part-time basis. Beyond a certain distance from the city, urban sprawl is reduced and most of the land is occupied by full-time farmers.

The actual pattern of land-use zones between Hamilton and Lake Erie was derived from an analysis of assessment data within the five categories of land use. The proportion of total assessment in each category was related to distance from the city by means of linear regression analysis. In the case of land used and owned by full-time farmers, residential land, commercial or industrial land, and vacant or idle land, a high degree of correlation indicated that the relationship between the portion of assessment accounted for, and distance from the city was linear in each instance. (Figure 11.3 shows the land-use distance relationships determined by the analysis and figure 11.4 depicts the land-use zones.) Residential assessment accounted for the highest proportion of total assessment close to the city and diminished outwards to zero at a distance of 46 km (29 mi). Full-time farming accounted for approximately 25 per cent of the total

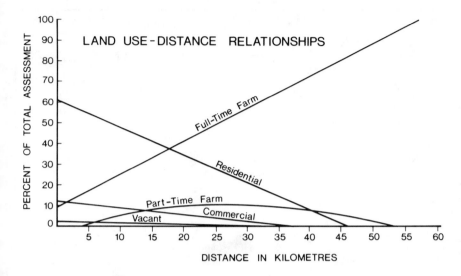

Figure 11.3 Change in land use with distance from Hamilton

assessment near the city boundary and was above 80 per cent at the 44-km (27-mi) point. Part-time farming reached a maximum of just over 10 per cent between 22 and 33 km (14 and 21 mi) from the city. Industrial assessment accounted for 10 per cent of the total at the city limits and decreases continuously to the 37-km (23-mi) point. Vacant land occupied 2 per cent near the city and decreased to almost zero at 40 km (25 mi).

The predicted limit of residential development was approximately 32 km (20 mi) in theory, whereas the actual development extended 45 km (28 mi). The discrepancy appears to be related to two factors. Some residents work in the local area and live close to their workplace. Other variation is explained by apparently irrational decisions. Residents beyond the zone of predicted residence are willing to pay an extra price for the aesthetic and social benefits derived from rural living conditions.

Data from questionnaires revealed insight into the motivation for migration to the countryside. Few people consciously realized the economics of rural residence. Most worked in Hamilton but did not consider the cost of commuting before deciding to move from the city. The two types of rural non-farm residents are: those with a former rural or small-town background who prefer rural to urban living, and those who migrate for economic reasons even though they may not understand this aspect accurately.

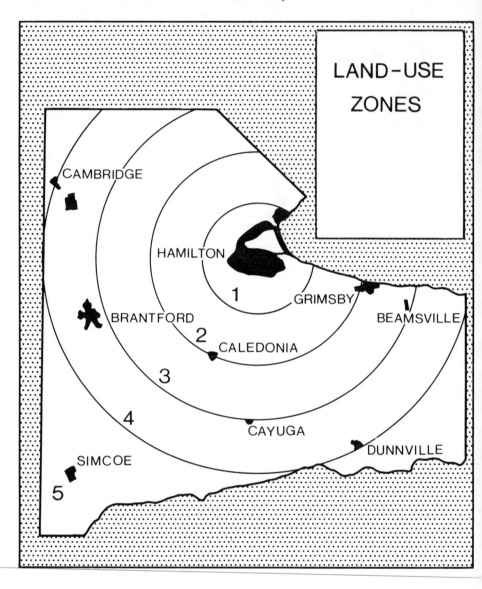

Figure 11.4 Land-use zones in the rural Hamilton area

TABLE 11.3
Criteria for determination of land-use zones

Zone 1: That area less than 11 km (7 mi) from the city centre but outside the corporate limits

More than 46% residential	More than 2% vacant or idle
Less than 7% full-time farm	More than 8% commercial
Less than 27% part-time farm	

Zone 2: That area between 11 and 22 km (7–14 mi) from the city centre

46%–32% residential	2%–1% vacant or idle
27%–45% full-time farm	8%–5% commercial
7%–10% part-time farm	

Zone 3: That area between 22 and 33 km (14–21 mi) from the city centre

32%–17% residential	Less than 1% vacant
45%–62% full-time farm	5%–1% commercial
More than 10% part-time farm	

Zone 4: That area between 33 and 44 km (21–28 mi) from the city centre

17%–1% residential	Less than 1% vacant or idle
62%–80% full-time farm	Less than 1% commercial
10%–6% part-time farm	

Zone 5: That area located at more than 44 km (28 mi) from the city centre

| More than 80% full-tme farm | Less than 6% part-time farm |

One may conclude that land use in the rural-urban fringe indicates a rather specific pattern which appears to be related to economic criteria and distance from the city. It is obvious that agriculture cannot compete with housing in an open market. As the costs of commuting increase, residential development in rural areas should decline but will not disappear unless more stringent zoning regulations are enacted.

The rural-urban fringe in some respects is a zone in which land-use changes are similar to those which occur in the soil (figure 11.5). The soil is a dynamic zone of the regolith that has changed gradually over time because of contact with the atmosphere and with plants and animals. The rural-urban fringe is a dynamic zone that undergoes change because of contact with the city. Beneath the soil is the relatively unchanged parent material, while at a critical distance from the city is the stable rural community. The depth of the soil varies with latitude while the extent of the fringe varies with the size of the city. The soil can be destroyed by poor agricultural practices while the rural-urban fringe can meet a similar fate if proper land-development procedures are not adopted.

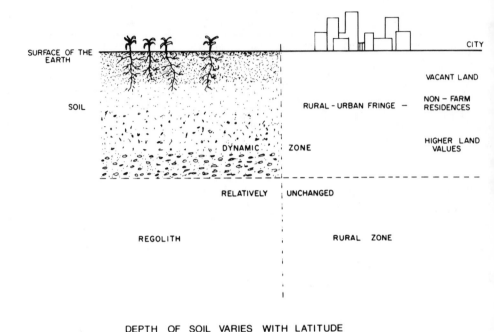

Figure 11.5 Land-use zones compared with soil horizons at the rural-urban fringe

THE NIAGARA FRUIT BELT

This account of land use would not be complete without a reference to the problems of the Niagara fruit belt. The loss of prime land has been an issue of great public concern since the declining area of the unique fruit land was reported in the research by Krueger (1959).

The bulk of the fruit being grown in the Niagara Peninsula comes from the north of the escarpment, the gentle Vinemount moraines immediately to the south of it, and the Fonthill kame. These areas have considerable acreages of sandy loam soil that are particularly well adapted to the production of peaches and sweet cherries, the tree fruits with the most exacting soil requirements. The well-drained deeper sandy loams have been referred to as the tender-fruit soils. Many are derived from the deltaic sands that were laid down on the Lake Ontario plain in the former glacial Lake Iroquois and from the fluvio-glacial kame deposits and others of the Fonthill area.

The climate of the lowland to the north of the escarpment is the most favourable in Canada for the growth of peaches and sweet cherries. Proximity to the moderating influence of Lake Ontario is a key factor since it reduces the probability of damage from severe winter temperatures and from spring frosts. The region has a sufficiently long growing season, adequate total heat, and one of the longest frost-free periods in Canada. The area benefits from the absence of high-velocity winds and a relative freedom from hail storms when the fruit is near maturity. Few areas on the continent possess such an ideal combination of soils and climate for the production of tender tree fruits.

Because of the fruit belt's favourable location in Southern Ontario in the industrialized Golden Horseshoe, and its good transport connections, it is also well suited to non-farm types of development. The die was cast with respect to the loss of land in the Niagara fruit belt when the Queen Elizabeth Way was built in the 1930s, for its construction marked the beginning of urban sprawl that was rampant until more stringent planning regulations were introduced in the 1970s. During this period, thousands of hectares of prime fruit-producing land were lost.

Research into the loss of farmland in Niagara is complex because the underlying causes vary greatly from area to area and from period to period. The individual farm enterprise responds to a multitude of factors, some of which are environmental and some socio-economic, while others relate to national or international political and economic conditions. All of these controls are commonly in a constant state of flux. To add to the uncertainties, the vagaries of the weather may bring about huge surpluses of fruit in one year and insufficient quantities to meet market demands in another.

Keeping the balance between preservation of this unique resource while permitting some urban development is the crux of the problem. One may sympathize with the city and regional politicians who are committed to sponsoring industrial growth and the provision of the related housing and services as a means of providing jobs and of reducing the tax rate for the beleaguered home owners. But conservationists succeeded in delaying the servicing of areas in Stoney Creek for industrial development when this was desperately needed to maintain economic growth in the Regional Municipality of Hamilton-Wentworth.

A realistic and rational solution to the problem of conflicting land uses in the Niagara fruit belt must still be sought. Problems affecting fruit production in Niagara relate to the cost-price squeeze, which means that production costs have been increasing at a faster rate than the selling price for the produce. Related issues are labour costs and availability, the need to invest large sums of money in mechanized equipment at a time when interest rates are prohibitively high, the problem of markets, foreign competition, and the increases in land prices and taxes for services not required for fruit production.

Fruit growers, like other entrepreneurs, do not want to be restricted in selling their land. The fruit industry has adjusted surprisingly well to changing economic conditions. With some protection from foreign competition and some relief from mounting assessment and the costs of borrowing, the industry will remain viable for many decades. In spite of losses of large areas of fruit land since 1951, the total production and value of fruit have increased.

It is my opinion that although zoning of land for agricultural use may be the most effective means of preserving the fruit land, it does not necessarily guarantee that the land will be used for fruit production. The industry will only survive provided the production of fruit continues to be a profitable enterprise. Therefore, the critical need is to introduce whatever measures are necessary to ensure that the industry remains economically viable. This calls for a co-operative effort on the part of the three levels of government: federal, provincial, and municipal.

Furthermore, it is my contention that zoning has an inbuilt injustice since it means that a property owner who happens to be located on tender-fruit soil forfeits his right to sell, while someone a short distance away on less productive land can benefit from the sale of his property to a developer. While it is is not feasible to compensate all farmers in Ontario whose land is zoned for agricultural use only, this seems to be a reasonable policy for the unique Niagara region (Reeds, 1969).

FUTURE PROSPECTS

Agriculture is in a transitional phase and at an economic crossroads. Adjustments to technological change are critical for the future of the industry. Mounting costs of production and too-low prices for produce have thrown many farmers into bankruptcy. Food prices must be permitted to rise in order to keep the industry solvent. The problems of loss of farmland and the loss of topsoil are unsolved. Agriculture needs to be given a higher priority in the planning process and ecological farming may be the answer to soil erosion and degradation. Monocropping practices and excessive use of synthetic fertilizers, pesticides, and herbicides have brought about a deterioration in soil quality.

It appears that mechanization will continue to expand, especially in the horticultural types of production. Increasing amounts of capital and credit will be required to finance the more highly mechanized operations. Land rental may become more important as a means of expansion and corporate farms may tend to replace the traditional family farm. Part-time farming will continue to increase, especially in the vicinity of the larger cities. Policies will be needed to assist the younger, marginal, low-income farm operator to transfer to other occupations. Increased demands for recreational open space and for all the services required to accommodate the expanding cities and industry will continue to put pressure on the food-producing land.

Farm operators will have to become more skilled managers and be better acquainted with marketing practices and research in order that they may be able to cope with the more sophisticated management tools and services. Computers, now used by fewer than 5 per cent of Ontario farmers, will become commonplace by the turn of the century. On-farm computers will enable the operator to improve productivity and management. Energy may be produced largely on the farm. If the greenhouse effect, due to the accumulation of carbon dioxide in the atmosphere, brings about an increase in temperature, the growing season will lengthen, permitting a wider choice of crops and new prospects for the farmers of the Hamilton-Wentworth and Niagara regions.

REFERENCES

Beesley, K.S., and Cocklin, C. 1982. Perspectives on the rural-urban fringe. *Occasional Papers in Geography* 2, Department of Geography, University of Guelph

Krueger, R.R. 1959. Changing land-use patterns in the Niagara fruit belt. *Transactions of the Royal Canadian Institute*, 32, Part 2, no. 67

– 1984. Where have all the fruitlands gone? *University of Waterloo Courier*, September Issue. Waterloo

Ontario Ministry of Agriculture and Food. 1965. Ontario Soil Survey Report no. 32, 'Wentworth County'

Reeds, L.G. 1969. *Niagara Region Fruit Belt Research Report*. Regional Development Branch, Ontario Ministry of Treasury and Economics, Toronto

Roberts, S. 1983. Agricultural Change in the Western Lake Ontario Region. MA thesis, Department of Geography, McMaster University, Hamilton

Rodd, R.S. 1976. The crisis of agricultural land in the Ontario countryside. *Plan Canada* 16: 160-70

CHAPTER 12

Social welfare
in the city

MICHAEL J. DEAR

As the numbers of hungry and homeless in Canada continue to grow, it is essential to remember how important cities are in helping disadvantaged and disabled peoples. In this essay, the remarkable concentration of social services in downtown Hamilton is examined. This concentration is associated with an extensive population of 'service-dependent' people. The geographical clustering of service facilities and people in need defines a distinct 'ghetto,' which has been a remarkably persistent feature of the city's geography since the early nineteenth century.

THE GEOGRAPHY OF SOCIAL SERVICES IN HAMILTON

The geographical distribution of social services in Hamilton is a reflection of the historical growth of the city (figure 12.1). Particularly significant in determining the historical distribution is the pattern of hospital provision in the city. Hospitals have usually been built relatively close to concentrations of population in Hamilton. Hamilton General Hospital, for instance, began life right in the core area and then moved (in 1882) to a more peripheral inner-city location on Barton Street. Only Hamilton Psychiatric Hospital (opened in 1876) possessed what could be called a suburban or semi-rural site. It was isolated upon the Mountain, away from the centre of population. After the Second World War, the expansion of population on the Mountain engulfed the psychiatric hospital, and led to the establishment of two new hospitals to serve the suburban population. Most recently, another hospital has been established on the west side of the city in association with McMaster University, reflecting the westward growth of population.

Hospitals are significant in that they generate ancillary health- and welfare-related activities in their vicinity. Not surprisingly, clusters of

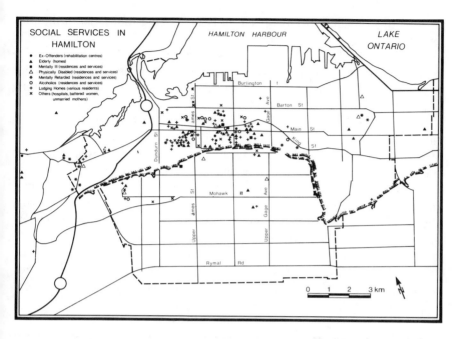

Figure 12.1 Location of social services in Hamilton

social services are located around the city's hospitals (figure 12.1). These include facilities for the elderly, ex-offenders, mentally disabled, mentally retarded, physically disabled, and alcoholics. The history of the growth of these facilities has been one of a steady intensification of a geographical pattern initiated in the nineteenth century. Two major clusters of service facilities exist on either side of the downtown business district, one of which is composed primarily of lodging homes. The development of the lodging-home district illustrates the self-reinforcing process of facility concentration characteristic of other types of services situated in the service-dependent ghetto.

Prior to the mid-1970s, lodging homes were essentially an unlicensed business in Hamilton. They were a private-sector phenomenon, mainly devoted to providing accommodation for the poorer inhabitants. However, the advent of 'deinstitutionalization' changed all that. This was the move away from institution-based care to community-based care, affecting many groups including the mentally ill, retarded, and physically disabled. Lodging homes grew in number throughout the late 1970s, and were increasingly involved in providing shelter for the discharged service-dependents.

This in itself might not have been an issue, except that most of the burgeoning lodging-home industry was concentrated in two wards close to the city centre (figure 12.2). This was an area where large houses were available for conversion to homes close to two of the city's five major hospitals. It was also an area where residents were relatively accepting of the newcomers – at least initially.

By 1976, there were 33 licensed lodging homes in Hamilton, plus an unknown number of unlicensed homes. The number had grown to 91 by 1980, and by 1984 had levelled off at 89 homes. During this period an increasing number of Hamilton's lodging homes entered into an agreement with the Regional Municipality of Hamilton-Wentworth to act as hostels under the General Welfare Assistance Act. Under the terms of this act, the operators provide more extensive services (e.g., 24-hour supervision) for 'special needs' clients. The lodging homes which participate in this program are private businesses, owned and usually run by their operators.

The contract lodging homes are funded through the regional municipality's Social Services department at a given per-diem rate. The rate is

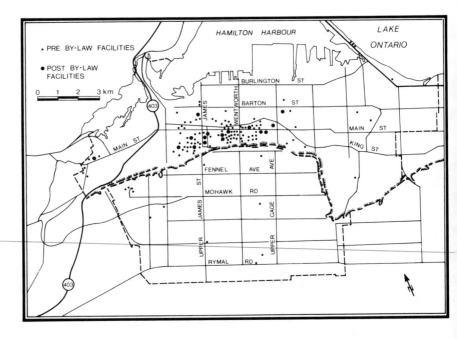

Figure 12.2 Location of lodging homes in Hamilton

set annually by the regional Social Services committee and is funded 80 per cent by the province and 20 per cent by the region. The province sets a ceiling above which the per-diem rate cannot rise if the 80 per cent funding is to continue. In 1979 the per-diem rate in Hamilton-Wentworth was $10.50 and the provincial ceiling was $13.00. (The local per-diem rate has been increased annually to keep pace with inflation.) A comparison of per-diem rates in various residential care facilities suggests that the lodging-homes option is very attractive from a cost point of view. It is by far the cheapest form of residential care, since it offers only a minimum of service and supervision. Thus, from the provincial government's viewpoint, a considerable cost saving was achieved by placing clients in lodging homes (at a time when the Ministry of Community and Social Services was experiencing severe cutbacks). At the same time, lodging-home operators recognize the opportunity for a steady, profit-making operation funded by the public sector.

The rise of the lodging-home industry in Hamilton should be viewed as a direct consequence of the 'restructuring' of the welfare state, and has taken the specific form of a less expensive type of residential care provided increasingly by the private sector. The state reduced the level of care at institutions, and simultaneously developed policies to incorporate highly fragmented and market-specific privatized or voluntary services. These political and legislative changes enabled the lodging-home industry in the core areas of Hamilton to develop as a free-market response. The marriage of state and private enterprise provided substantial fiscal savings for the welfare state and ever-increasing markets and profits for private entrepreneurs.

The effect of restructuring on client populations can be observed in the socio-demographic structure of the lodging-home population. A survey of 717 residents conducted in 1980 showed that more than half of them had previously been resident in psychiatric hospitals; 14 per cent had come from general hospitals (usually elderly people); 9 per cent had been in trouble with the law; and 6 per cent had been referred from organizations dealing with alcoholism. Only 12 per cent had moved directly from private addresses to lodging homes. It is interesting to note that 26 per cent of this population were not resident in Hamilton before entering the program. More than 30 per cent of the residents were over sixty-five years of age, and about 34 per cent were under twenty-four years old. The remaining 36 per cent were evenly spread across the intervening age category (twenty-five to sixty-four). At least 77 per cent of the lodging-home residents depended upon direct government transfers as a source of income. Some of the 22 per cent whose income was unspecified may also depend upon government transfers but did not declare this for various reasons. Almost half of the people existed on Family Benefits Assistance (FBA) or Gains-D programs, indicating that they were either part of the long-term welfare population or had a permanent disability. Another 25 per cent of the population existed on some form of government pension, reflecting the sizable number of elderly people who live in these homes. Most of these people live on incomes which are below the poverty line and which have been decreasing in purchasing power in recent years. Generally the cost of food and shelter at the lodging home is deducted from the welfare recipients' incomes so that they are left with a disposable income of approximately $60 per month (1984).

During 1975 and 1976, one local alderman noticed that he was receiving an increasing number of phone calls from irate constituents. They were angered and bewildered by what they perceived as frequent new openings of lodging homes in two wards close to the city centre. The issue was brought into focus by a tangential occurrence in another neighbourhood. There, potential operators of a doughnut shop had applied for a licence. Following usual procedures, the application was advertised, and inspired an intense community opposition. As a consequence, the application was refused by the city. The operator appealed the decision, and the court upheld the appeal. It was found that an application for an operator's licence could not be refused simply on the grounds that some neighbours objected to the operation.

This seemingly innocuous appeal had a profound effect. It caused an inquiry into the city's licensing procedures. As a result, a formal licensing committee was established; its task was to rule on matters of fact and not to make political judgments. Specifically, the advertising of licensing applications was stopped, and the fulfilment of licensing standards (e.g., health, safety, and fire regulations) became a purely technical matter. This licensing inquiry also uncovered much information about illegal operations in the city, and about the quality of care/accommodation in lodging homes.

The emergent community opposition to neighbourhood 'saturation' by lodging homes, together with concerns about licensing and standards, found a combined voice in the 1977 *Report on Lodging Homes, Halfway Houses and Nursing Homes*. Prepared by Alderman Brian Hinkley, this deliberately provocative report drew attention to the problems of the service-dependent in lodging homes. The subsequent fire-storm of public concern led to a 1978 conference on community residential services, which concluded that deinstitutionalization was a good thing but that ghetto-ization/saturation was not.

A new city by-law was drafted in June 1978. It included much tighter definitions and licensing procedures. A 'lodging house' was defined as: 'a dwelling in which four or more persons are lodging for gain, with or without food and without separate cooking facilities, by the week or more than a week and which is licensed as a lodging house.' A 'residential care facility' (RCF) was defined as: 'a fully detached residential building oc-cupied wholly by a maximum number of supervised residents ... on the premises as a group because of social, emotional, mental or physical handicap or personal distress for the purpose of achieving well being.'

The by-law also introduced a distance-spacing requirement; no RCF may locate closer than 180 m (590 ft) from lot line to lot line of another facility. Capacity requirements were also set out. Finally RCFs were class-ified as 'permitted uses' in all residential and commercial zones of the city. This condition was introduced because it was the specific intent of the legislation that lodging homes be accepted throughout the city (even though 'fair-share' provisions did not appear in the by-law). Previously the absence of the 'permitted use' designation had frequently caused an application for zoning variance to be made; this application tended to be a significant factor in alerting community opposition.

Appeals, delays, and modification to the legislation ultimately post-poned its approval by the Ontario Municipal Board until 1980. The law became operative in 1981. By that time, however, the number of lodging homes in the city had peaked. The form of the ghetto had been established (figure 12.2).

In recent years, growth in the lodging-home industry has slowed. In January 1984, the vacancy rate in lodging-home beds was 10 per cent (comparable rates exist in Ottawa, a city of equivalent size). The region will now license new homes only if an existing home closes or a demon-strable need can be shown. Notwithstanding these strictures, new homes continue to be opened; to the end of 1984 seventeen had been opened since the passage of the by-law in 1981. However, new 'births' are also accompanied by 'deaths' in the industry, and some homes are going out of business. There is some evidence to suggest that residents are 'filtering up' into the newer and better-quality homes.

Despite the new by-law, the new lodging homes continue to adopt a ghetto location (figure 12.2). This is largely because the choice of home location remains in the hands of the operator, once formal licensing and distance-spacing requirements have been met. In 1983, for example, seven out of the nine new homes opened were located in the core area. This is a reflection of operators' preference to be close to the centre of demand, as well as the increasing difficulty of finding properties suitable for conversion outside the core. Three smaller nursing homes in the region have recently been converted to lodging homes. Hence, the net effect of the new by-law has been to reduce the crowding of facilities; but it has not had a significant effect on the core-area concentration of facilities. Instead, there has been an increase in rezoning to increase capacity in existing facilities and spot zoning to allow specific variances in the distance-spacing requirements.

Despite the continuing ghettoization, the more liberal and precise by-law seems to have had the effect of appeasing community opposition, at least temporarily. Residents in the most impacted zones now seem happier with the situation, although they need no longer be informed of a home's opening. Another factor in the present truce is that home operators have increased their public-relations outreach, and are more conscious of the need to communicate with their host communities.

However, Alderman Hinkley has warned that if the intent of the by-law is ignored, then the ghettoization issue is likely to resurface and undermine the lodging-home program. Clearly, while operators retain discretion over locational choice, the ghetto is unlikely to be dismantled. But the provincial government is searching for ways to undertake a more directive role in facility siting. Several suburban jurisdictions around Hamilton are preparing their own by-laws in anticipation of expansion in the lodging-home industry. Meanwhile, the growth of lodging homes, group homes, and other facilities into suburban jurisdictions in Toronto is causing intense opposition in some traditionally exclusionary neighbourhoods. The future of the service-dependent ghetto in Hamilton is, at least for the present, highly uncertain.

THE GEOGRAPHICAL DISTRIBUTION OF SERVICE-DEPENDENTS IN HAMILTON

The concentration of helping agencies in downtown Hamilton has attracted a large population in need of their services. The clustering of five important client groups is shown in figure 12.3. These are the mentally ill, mentally retarded, physically disabled, elderly, and probationers and parolees. For these groups, a downtown location is a positive thing; the inner city has become part of their mechanism for coping. There they can meet friends, obtain help from professionals, and find accommodation.

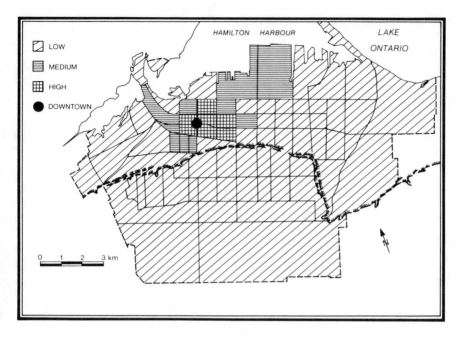

Figure 12.3 Distribution of social service demand in Hamilton

The particular case of psychiatric patients is illustrative of the significance of the downtown location. By far the largest proportion of admissions to Hamilton Psychiatric Hospital (HPH) come from six census tracts
close to the downtown (figure 12.4). The pattern of discharge is even more
clearly defined (figure 12.5). In many ways, the inner city acts as a 'reservoir' of admissions to and discharges from HPH. How do these people
end up in the inner-city 'ghetto'?

A group of ex-psychiatric patients awaiting discharge from HPH have
hopes and fears which are similar to those of many groups planning a
new move. They have common concerns over housing and neighbourhood
quality, over their ability to find work, and to make ends meet. They are
not ignorant of the hazards of coping in a new environment, and have
begun to formulate plans, however tentatively. Some of these plans reflect
a depressingly accurate realism: everyone expects to receive an income
which is below the poverty line; many expect to be isolated from family
and friends. Other expectations are about to be dashed upon discharge:
few thought they would have problems with housing; their clear desire to
be employed becomes overridden by worries about bleak job prospects;

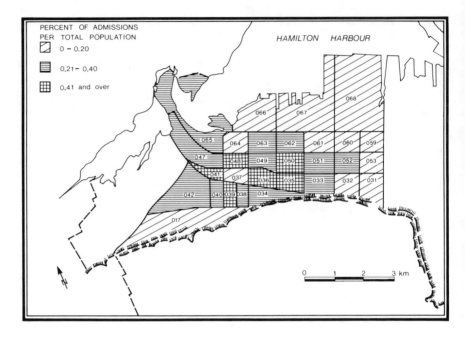

Figure 12.4 Distribution of admissions to Hamilton Psychiatric Hospital

and the anticipated active social and recreational program does not materialize through lack of access to such programs.

Once in the community, the everyday lives of ex-psychiatric patients reveal an incredible diversity, reflecting their varying success in adjusting to the outside world. One ex-patient described his typical day as follows: 'During the week, I get up – go to work from 9 to 5, sometimes grocery shop on my way home, cook supper, and then in the evening I either go swimming or have a game of ball with my friends. Once a week I come to —— for my appointment. I play my guitar whenever I have a chance.' He seems to be coping very well in the community. Although still dependent on a psychiatric service, he creates a clear picture of his ability to hold a job, cook, shop independently, and involve himself in recreational activities. He also indicates that he is socially involved with others.

Contrast this with an ex-patient who seems to be having trouble coping with community life. Although he is not using a follow-up service regularly, it appears that this person has been left to fend for himself. Many of his needs seem to be unmet. He does not appear prepared to face the day-to-day problems and routines that life entails. Thus: 'I have to get up at

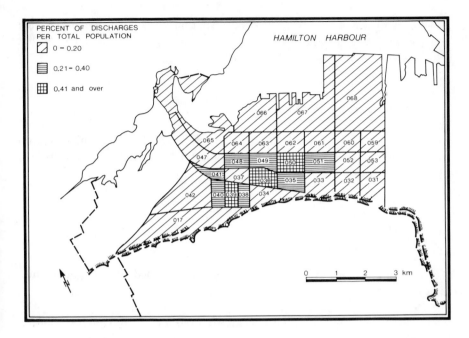

Figure 12.5 Distribution of patients discharged from HPH

7:30 for breakfast, I go back to bed until lunch, I watch T.V. or walk in the afternoon, and I go to bed after dinner.'

The everyday life of the ex-patient without a job tends to be poorly structured, i.e., lacking in any clear routine or purpose. Weekends are spent in much the same way as weekdays, although activity levels drop in response to the diminished opportunities for organized social interaction. Specific problems encountered by the group relate to unemployment, housing, psychiatric and social services, and loneliness.

The chronic psychiatric patient is, in effect, consigned to permanent unemployment. A few subsidized jobs exist; they provide activity, some dignity and a sense of self-worth, which are just as important as the small amounts of money received.

I've worked a long time building up my credentials and now even though I feel I fit in, people still turn me down for employment.

My social worker said [I] wouldn't be good for the job – shipping and receiving with a plumbing outfit: [my] social worker didn't want any failures. He wants me to go back to school but I want to work. I don't think spelling is required for the job I want. I would rather work and try it out. [My] social worker feels that because I was in the hospital, I'm not fit to work.

I think jobs are the key to better mental health.

I would feel useful if I had a job.

Poverty imposes severe restrictions on choice of housing, which is often reduced to selecting one of a number of indifferent single-room units in a downtown area.

It's hard to find a place to live – people avoid me.

[I] don't like the neighbourhood: [it's a] slum, people are always fighting: [the] area is really rundown.

I couldn't find a place to live, —— Housing Agency turned me down three times in a row.

Poverty-stricken, and unable to afford anything but the cheapest housing, the ex-patient might anticipate support from the professional aftercare support network. This, however, often fails by doing too much or too little.

I feel sometimes I need psychiatric attention but I don't know where to go to get it. I'm afraid they'll just put me in the hospital.

The psychiatric attention I get is not adequate. I feel funny after medication. They only cut down on medication; they don't explain anything; they keep secrets.

The attention I get is almost excessive at times.

I feel I'm being taken advantage of because of my situation. I took 35 or 40 shock treatments. I cooperated with the doctors when I didn't know what they were doing to me. People always see me as a psychiatric patient when I want to be normal and respectable.

The drift to the ghetto in search of informal support networks is often accelerated by the professional care-givers. In a follow-up study of 495 patients discharged from HPH between 1 April 1978 and 31 March 1979, it was found that 70 per cent were actually discharged to destinations in the core area. The discharged population was confined predominantly to census tracts close to the central business district and included a significant number (12 per cent) of people who had lived outside of Hamilton prior to admission to HPH. (Many outsiders seem to remain in Hamilton after discharge.)

The social networks of the discharged mentally ill in Hamilton tend to be much smaller in size than those of the non-disabled. The mentally ill have fewer ties with kin, fewer friends, less interaction with friends and relatives, and fewer different sources of friends.

I feel terribly lonely, like I'm forgotten by everyone.

I don't like going home to my family because as soon as I get there, everyone goes out and leaves me there alone.

Since I was ill, I've lost out on the finer things in life. I used to enjoy going to the opera, the ballet, ... but haven't been able to in a long time.

I'm not accepted in the community because of my sickness.

In summary, the experience of discharge can effectively dash the hopes and optimism of many patients as they re-enter the community. They face severely limited (usually downtown) housing opportunities; they are frequently referred upon discharge to core-area accommodations and services which are often found to be unsatisfactory and ineffective; and they are forced to turn inward in the face of diminished social networks. In their search for 'community,' those ex-patients (who have not already been referred there) gravitate towards the inner city. It is almost self-evident that the impetus behind ghettoization is the search for a wider support network. The inner city has become a coping mechanism where ex-patients can find help in the search for jobs and homes, can locate other support facilities, begin or renew friendships, start self-help groups, and operate newsletters. In the absence of any better alternative the ghetto is functional for Hamilton's ex-patients. It is a spatially limited zone where access to different kinds of support is made possible through geographical proximity.

NOTE

This chapter is an extract from M.J. Dear and J.R. Wolch, *Landscapes of Despair: From Deinstitutionalization to Homelessness* (London: Polity Press/Blackwell, 1986), and is partly based on M.J. Dear, L. Bayne, A. Boyd, E. Callaghan, and E. Goldstein, *Coping in the Community: The Needs of Ex-psychiatric Patients* (Hamilton: Canadian Mental Health Association, 1980).

CHAPTER 13

The changing competitive position of the Hamilton steel industry

W.P. ANDERSON

The iron and steel industry has been the driving force in Hamilton's economy throughout the twentieth century. In recent years, Hamilton's dependence on this sector as a source of manufacturing employment has actually increased. This dependence raises serious questions about the future. The steel industry is currently characterized by declining employment and unused capacity, not only in Hamilton but throughout North America and Europe. In the face of stagnant demand for steel products, no major growth may be expected in the near future.

The technology of the iron and steel industry has improved markedly over the past thirty years. Improved methods of iron-ore reduction, steelmaking, and the casting and rolling of steel products have increased the productivity of steelworkers in Hamilton as elsewhere. Ironically, in an industry that has ceased growing, increased productivity can only lead to a decline in employment. Hamilton's steelworks must continuously modernize to remain competitive, and jobs are likely to be lost in the process.

If the current levels of employment are to be maintained, the steelworks of Hamilton must compete more successfully with other steel production in Canada. In other words, they must take a larger slice of a pie whose size appears to be fixed. In the past, Hamilton's competitive record has been outstanding. Its share of Canadian steel production increased consistently, reaching over 70 per cent by the mid-1970s. However, the trend may be reversing. The Steel Company of Canada (Stelco), the largest steel producer in Canada and by far the largest private employer in Hamilton, recently built one of the world's most efficient steelworks. Stelco bypassed Hamilton harbour which had previously been the location of all its integrated steelworks, and chose instead to site the plant on the north shore of Lake Erie at Nanticoke. While the construction of this new plant is a positive step for the Canadian steel industry as a whole, it represents a decline in the relative importance of Hamilton in the industry.

In this essay, I will examine the current competitive position of Hamilton in the steel industry. In order to do this it is necessary to look first at changes that are affecting the steel industry throughout the world. These changes involve technological developments, the growth rate of market demand, and increasing international competition. Within the context of these changes, it is possible to examine the condition of the Canadian steel industry and the competitive position of Hamilton's steelworks within the Canadian steel industry.

CURRENT TRENDS IN THE STEEL INDUSTRY

Technological Trends

Technological improvements in the iron and steel industry are designed to reduce production costs by reducing the amounts of materials, fuels, and labour inputs needed to produce a tonne of steel. Materials and fuels are saved by designing machinery that operates more efficiently and thus wastes less of these substances. Labour inputs are reduced by designing machinery that requires less maintenance, may be operated by fewer people, or takes over tasks once performed by steelworkers. Naturally, machinery of this type is only practical if it can produce savings in materials, energy, and labour costs which outweigh its own cost, which is called a 'capital cost.'

When a firm installs new steel-producing capacity, it naturally purchases capital which embodies the latest production technology. As the useful life of capital in the steel industry is quite long, the firm will be 'locked in' to that technology for years to come. During these years, other firms may install capital which embodies even newer technological improvements. Thus, as a firm's capital stock ages, other firms may gain a competitive advantage over it.

Under circumstances of gradual technological change, this type of competitive advantage is only temporary. As each firm in the industry undertakes periodic modernization, the advantage shifts back and forth among the competitors. However, when technological improvements which imply dramatic production cost savings are introduced, some firms may find themselves at a permanent disadvantage. In such cases, the firm will experience declining profitability, which may reduce the confidence of investors and banks in its viability, and thereby reduce its ability to borrow for needed capital modernization. Thus, the firm finds itself in a kind of technological trap – unable to compete with the stock of machinery it has, and unable to purchase the new machinery it needs to become competitive.

Over the past thirty years, a number of technological innovations which result in sufficiently large cost savings to produce just this type of situation have been introduced in the steel industry. In fact, this type of competitive disadvantage has affected entire regions and countries. For example, the steel industries of the United States made large investments in the 1950s in technologies that were soon to become obsolete, placing them at a competitive disadvantage in world markets from which they have yet to recover (Barnett and Schorsch, 1983).

In order to consider the implications of specific technological improvements, it is necessary to have at least a rough understanding of the way iron ore is transformed into useful steel products. The steel-production process may be divided into four distinct but interrelated stages: iron-making, steel-making, production of semi-finished shapes, and finishing.

The iron-making stage actually involves two sub-stages: materials preparation and iron reduction. The principal materials used are coal and iron ore. The coal is pre-burned in a battery of ovens in order to remove its most volatile elements. The result is a slow-burning fuel called coke. Iron ore is treated in order to increase its ferrous content and to transform it into a pellet-like consistency. Coke ovens are generally located at the steel plant while iron-treatment facilities may be located either at the steel plant or at the mine site.

The iron and coke, along with a third material, limestone, are introduced into a large vessel called a blast furnace. Inside the furnace a blast of hot air ignites the coke and the resulting heat reduces the iron ore to a liquid. The limestone acts as a flux to remove impurities so that relatively pure molten iron may be tapped out of the bottom of the furnace.

A wide variety of technological innovations have been introduced in both materials preparation and iron-ore reduction. Most of these have provided only marginal costs savings. The most significant technological trend in the iron-making stage is the construction of increasingly large blast furnaces. Larger furnaces make more efficient use of the heat energy contained in the coke and make it possible to extract a higher proportion of the ferrous content in the iron ore. Since it is the iron-making stage in which the greatest cost advantages of large-scale operation are gained, the minimum efficient size of steel plants is dictated by the minimum efficient size of blast furnaces.

The purpose of the second stage is to transform iron into steel, which is a more workable metal that may be tempered to a variety of hardnesses. This is achieved by heating the molten iron to over 2000°C. At this temperature, chemical reactions that reduce the carbon content of the metal occur. Steel scrap is also used as an input in the steel-making stage. The scrap may be purchased from salvage firms or obtained from cutting and trimming of steel in the later stages of production.

In the late 1950s, most North American steel was produced in the open hearth, which is a shallow vessel in which molten iron and scrap are heated together. In the early 1960s, the open hearth was displaced by the basic oxygen furnace. In the basic oxygen furnace, pure oxygen is blown into the molten material, causing chemical reactions to occur rapidly. The basic oxygen furnace has two major advantages over the open hearth. First, while the open hearth requires natural gas or distillate oil as a fuel to produce the required high temperatures, the basic oxygen furnace operates exclusively on the heat produced by its chemical reactions. Thus, an energy savings is achieved. Secondly, the basic oxygen furnace can make steel in a 'heat time' of less than one hour, as compared to six hours or more in the open hearth. These advantages made the open hearth obsolete by 1970, and put those firms still using open hearths at a tremendous disadvantage.

Both the open-hearth and basic oxygen-furnace steel-making methods are applicable only in integrated steelworks. Integrated steelworks are plants where iron-making and steel-making are carried out at the same site. Naturally, since both these methods use molten iron as an input, they must be located near a blast furnace. An increasing share of the world's steel is now being produced at non-integrated steelworks – that is, plants that do not include blast furnaces. These steelworks use a third type of steelmaking technology: the electric furnace.

The electric furnace produces molten steel exclusively from steel scrap. The scrap is literally melted down by an electric charge. This technology has been in existence since the early 1950s, but technological improvements throughout the 1960s and 1970s have led to its constantly increasing use. Since it does not require molten-iron input, steel firms employing the electric furnace are not locationally tied to sources of iron ore. Also, since it is not used in conjunction with a blast furnace, the minimum efficient size for steelworks using the electric furnace is much smaller than for integrated plants. For this reason, these steelworks are often called 'mini-mills.' The principal drawback of the electric furnace is that it is capable of producing only a limited range of steel products, though the range has been extended somewhat in recent years.

In the third stage of the steel-production process the steel is formed into semi-finished shapes, which are large pieces of cold steel that are later passed on to the finishing stage. There are three basic semi-finished shapes: blooms, which are square in cross-section and several feet long;

billets, which are shaped like blooms but are thinner, and slabs, which are broader and flatter in cross-section.

In older steelworks, the transformation of the molten steel in the steel-making vessel into semi-finished shapes requires a number of intermediate steps. First, the hot steel is poured into moulds where it cools to form large lumps of steel called ingots. The ingots, which weigh 15 to 20 tonnes (16.5 to 22 tons), are transported to a separate facility called a primary rolling mill. There they must be reheated so that they may be passed through high-pressure rollers to form blooms, billets, and slabs.

More recently, a process innovation called continuous casting has been introduced as an alternative method of producing semi-finished shapes. With continuous casting, the intermediate steps of ingot-production, re-heating, and primary rolling are bypassed entirely. Instead molten steel is poured directly into moulds to form the semi-finished shapes. Continuous casting has several important advantages. It saves energy by elim-inating the reheating of ingots and allows superior quality control. Also, it involves much lower capital costs than the huge primary rolling mill that it replaces. However, because the continuous-casting process re-quires a regular throughput of hot steel, it cannot be used in conjunction with an open hearth, which produces an irregular flow of steel because of its long heat time. For this reason, continuous-casting techniques can only be used in conjunction with a basic oxygen or electric steel-making furnace.

In the following stage of steel production, the semi-finished shapes are transformed into a variety of products in specialized rolling mills. Blooms are rolled into structural shapes, such as the beams used in the construc-tion of skyscrapers. Billets are rolled into rail, rods, and wire. Slabs are rolled into plates, such as those used in ship-building, and sheets, such as those used in automobile production. Treatments such as galvanizing and tin plating which improve the surface quality of the steel are also completed in this stage. At present, most electric-furnace 'mini-mills' can only produce billet-based products. Integrated steelworks may produce the entire range of steel products, although some specialize in a limited range of finished products.

Market Trends

Steel industries in industrial countries currently face a relatively flat mar-ket – that is, a market in which the demand for steel products is either stagnant or increasing only slightly. Most of these countries experienced a period of rapid increases in demand in the 1960s or 1970s, before settling

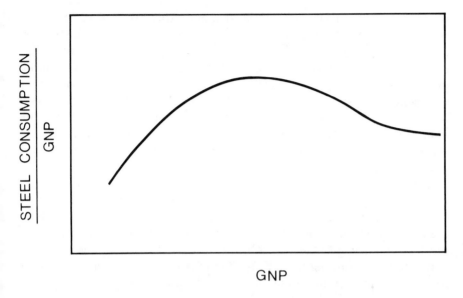

Figure 13.1 Schematic relationship between GNP and steel consumption

into a pattern where the rate of growth in steel consumption is considerably smaller than the rate of overall economic growth.

Figure 13.1 displays a relationship between steel consumption and overall economic growth which has been typical of western countries over the past forty years (Barnett and Schorsch, 1983: 38). The vertical axis measures the ratio of steel consumption to gross national production (GNP). This is essentially a measure of how 'steel intensive' the economy is. The horizontal axis measures GNP per capita, which is an indicator of economic progress. The early stages of economic development require major investments in infrastructure – transport systems, buildings and heavy machinery, etc. – which in turn require huge quantities of steel input. Thus, the ratio of steel consumption to GNP increases for a time. As the economy matures, however, the completion of major infrastructural projects, shifts in employment away from heavy industries, and the development of new materials which take the place of steel work to reduce the ratio of steel consumption to GNP. As long as this ratio is in decline, a relatively flat market for steel persists.

In addition to, and in part because of slow market growth, steel industries in North America and Europe are troubled by low ratios of capacity utilization. This means that a large part of the physical apparatus of steel production lays idle, at least temporarily. Several factors have contributed to this situation. First, many steel firms expected the healthy rate of market growth that they enjoyed prior to the mid-1970s to continue, so they invested in more facilities than were needed in the markets of the late 1970s and early 1980s. Second, in many countries, government investment in the steel industry kept many facilities which might otherwise have been closed operating at a fraction of their capacities. Third, the volatility of post-OPEC western economies dictates that in order to fulfil market demand during upswings, excess capacity must be maintained during slumps.

One result of this condition is a general reduction in the cost efficiency of steel plants. When a facility operates below its capacity, excessive capital costs are incurred and energy efficiency and labour productivity are reduced. Furthermore, when capacity utilization is low and unstable, the labour force is subject to job insecurity and frequent lay-offs. This situation is devastating not only for the lives of the steelworkers, but for to the economies of the communities in which they live.

International Competition

The international distribution of steel production has changed significantly during the post-war years. Output in the traditional steel powers, such as the United States and Germany, has stagnated and even declined. In the United Kingdom, steel production has shrunk by almost 50 per cent from its peak levels of the 1960s. Meanwhile, Japan has emerged as the world's leading exporter of steel products, and South Korea appears ready to follow in its footsteps.

Many developing countries, which were at one time customers for steel exports from Europe and North America, are currently developing their own industries. Some developing countries, such as Brazil, are self-sufficient in steel, while others, such as the Philippines, have identified growth in steel production as a major planning objective.

The decline in the dominance of western steel-making nations is partly result of the technological trends outlined above. For example, Japan was left with only a very small steel-production capacity at the end of the Second World War. While this meant that enormous investments were required to increase steel production, it also left the industry unencumbered by obsolete equipment. Also, technical advances have reduced the requirement for some classes of skilled labour, thus reducing the advantage of western countries over developing countries. Escalating labour costs and stagnant domestic markets also contributed to the decline.

The heightening of competition in the steel industry at a global scale reinforces the need for steel firms in the west to adopt the latest technological innovations in order to protect their traditional markets.

THE CANADIAN STEEL INDUSTRY

Origin and History

A number of factors contributed to the rapid growth of integrated steelworks in Canada in the late years of the nineteenth century and the early years of the twentieth century. The 'National Policy' of protection introduced by Prime Minister Macdonald in 1878 freed the blast furnaces and rolling mills of Ontario and Quebec from competition with larger U.S. firms. Some large firms, including one in Hamilton, were funded with initial investment from the United States. The opening up of Canada's west led to increasing demands for steel for the railroads and the manufacture of agricultural equipment. As a result, 'between 1901 and 1911, iron and steel and their manufacturers became Canada's foremost industry' (Kilbourn, 1960).

By this time, three major steel-producing centres were well established: Hamilton, Ontario; Sault Ste Marie, Ontario; and Sydney, Nova Scotia. Each of these centres had significant locational advantages for the transportation of iron and coal. Steelworks at Hamilton, at the extreme western end of Lake Ontario, originally made use of some local materials, but gained access to coal from Appalachia and iron ore from the Lake Superior region with the expansion of the Welland Canal in 1887. Sault Ste Marie is the location of a canal linking lakes Superior and Huron, and is therefore an easy place to bring iron ore and coal together. Sydney, Nova Scotia, is ideally located to exploit Nova Scotia coal and Newfoundland ore. While these three have remained the dominant steel-producing centres in Canada, Hamilton has overshadowed the other two, in part because its superior access to the large steel consumers located along the shores of Lake Ontario.

While the Canadian steel industry suffered stagnation and physical deterioration during the depression years of the 1930s, it rebounded and expanded in the 1940s as a result of the demand for steel products during the Second World War. To fulfil the needs of the mobilized economy, steel plants were expanded with government aid. After the war, the government continued to encourage investment by allowing accelerated depreciation schedules by which fixed assets could be written off in just two years. This gave firms making large investments a considerable tax benefit.

A major problem faced by steel firms in a relatively small market like Canada's is the inability to operate at a sufficiently large scale. Since economies of scale in steel production are great, an industry made up of relatively small plants is highly vulnerable to the encroachment of imports from larger, more efficient foreign plants. If every steel firm in Canada were to produce a full line of steel products, none would be able to produce any single product at sufficiently large scale to offer a competitive price. In order to prevent this from happening, individual firms have tacitly agreed among themselves to specialize in limited product lines (Barnett and Schorsch, 1983). So, for example, Dofasco in Hamilton specializes in flat rolled products, while Algoma in Sault Ste Marie specializes in structural shapes, rails, and plates. Only Stelco, the Canadian industry's undisputed giant, offers a truly broad line of steel products. While some might argue that this sort of market-sharing behaviour violates the principles of free trade and competition, others counter that it has been necessary in order to allow Canadian firms to operate at efficient scale.

The Comparative Performance of the Canadian Steel Industry

The recent performance of the Canadian steel industry may best be gauged by comparing it to that of the steel industries of other countries. Figure 13.2 shows the production record of the past twenty years for steel industries in Canada, the United States, the United Kingdom, West Germany, and Japan. Two of these countries, the United States and the United Kingdom, have suffered very significant absolute declines since the mid-1970s. West Germany, after healthy growth in the 1960s, has settled into a period of little or no growth which is consistent with the later stages of economic development. Even Japan, where the growth in production of the 1960s indicates one of the great success stories of industrial history, has entered a period of stagnation in steel production.

By comparison, the Canadian steel industry enjoyed moderate but consistent growth in the 1960s and 1970s. In retrospect, this type of growth appears almost ideal for the development of a steel industry. Consistent growth has made it possible for new technologies to be incorporated as the stock of capital was expanded. However, since there was no 'boom' period, the temptation for massive overexpansion of facilities was not present.

In table 13.1, indicators of the state of technology for steel industries in the same five countries are presented. It was mentioned earlier that the open-hearth steel-making process is generally considered to be obsolete. For this reason, a shift from this process to the basic oxygen and

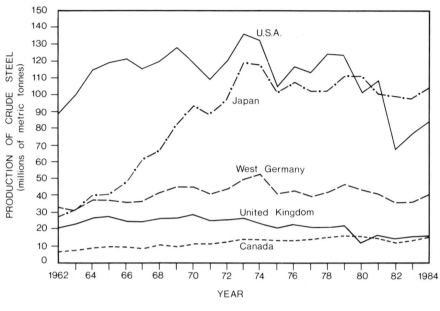

Figure 13.2 Steel production in the major producing countries, 1962-84

electric furnaces indicates technological improvement. The data for 1981 show that Canada and the United States have lagged behind their European and Asian competitors in this respect, with 14 and 11 per cent, respectively, of their outputs of crude steel coming from open hearths. It should be noted, however, that in Canada this percentage should be much lower by 1985 with the addition of additional basic oxygen-furnace capacity and the retirement of several open-hearth operations.

The percentage of steel continuously cast is another indicator of the state of production technologies. While Canada has surpassed the United States and the United Kingdom in this category, it still lags behind West Germany and Japan. Massive investments in continuous-casting equipment by two Canadian steel firms, Stelco and Dofasco, should bring this figure well over 50 per cent by the close of this decade.

Perhaps the most universally applied measure of technological proficiency is the productivity of labour, measured here as thousands of tonnes of crude steel per employee. Here Canada outperforms all of the countries listed except Japan, the acknowledged technological leader. This is indic-

TABLE 13.1
Comparative indicators of steel-production technology

		Canada	United States	United Kingdom	West Germany	Japan
Percentage breakdown by process						
Open hearth	1981	14.0	11.0	0.0	0.3	0.0
	1975	23.6	19.0	22.0	16.7	1.1
Electric	1981	25.7	28.3	32.4	15.9	24.8
	1975	20.3	19.4	27.6	12.6	16.4
Basic oxygen	1981	60.3	60.6	67.6	80.2	75.2
	1975	56.1	61.6	50.4	69.3	82.5
Per cent continuously cast	1981	33.2	21.6	31.8	53.6	70.6
	1975	12.8	7.3	8.9	24.2	27.2
Output per employee (tonnes $\times$ 10^3)	1981	270.4	255.3	162.2	216.7	377.7
Capacity utilization rate (%)	1981	76.2	78.3	61.4	61.3	n.a.

SOURCE: *The Iron and Steel Industry in 1981.* Paris: Organization for Economic Cooperation and Development 1983

ative not only of the quality of Canadian labour, but also of the labour-saving equipment installed by Canadian firms.

Canada's capacity utilization rate for 1981 is hardly admirable, with nearly a quarter of the available plant and machinery lying idle. However, capacity utilization in the United Kingdom and West Germany is far lower. This advantage is in large part a result of the steady, moderate expansion of steel-making capacity in Canada during the past two decades.

The Canadian steel industry's recent performance in international markets is heartening. In only a few years, Canada has changed from a net importer to a net exporter of steel. In 1974, steel imports were 2.6 million tonnes (2.9 million tons) against 1.1 million tonnes (1.2 million tons) of exports. By 1980, the situation was completely reversed, with 1.1million tonnes (1.2 million tons) of imports against 3.1 million tonnes (3.4million tons) of exports. While Canada is never likely to become a major exporter of steel, the fact that 20 per cent of production goes onto the world market speaks well for the competitiveness of the industry.

The industry also appears to be well able to defend itself against encroachment by foreign producers. A comparison with the United States is instructive in this regard. Because of a slump in demand, 1982 was a poor year for steel producers throughout the world. Table 13.2 shows the effect of this slump on domestic production and imports in the United States and Canada. In the United States, production dropped from its 1981 level by 37 per cent, while imports declined by only 16 per cent. In Canada, the reverse occurred. Domestic production declined by 18 per cent, while imports declined by 64 per cent. Thus, the Canadian steel industry was able to survive the slump by driving out imports, while in the United States it was the imports that were driving out American steel.

While the Canadian steel industry faces a somewhat uncertain future, its performance over the past two decades should not be underestimated. Its record of growth and capacity utilization and its performance in international markets are outstanding among western nations. On the negative side, Canada, like the United States, has been slow to adopt new production technologies. However, modernizations either completed in or planned for the 1980s should correct this situation.

HAMILTON'S ROLE IN THE CANADIAN STEEL INDUSTRY

Early Beginnings

Many factors contributed to the early development of the iron and steel industry in Hamilton. Four prominent ones are: 1 / the building of railways; 2 / federal and local government policies to encourage industrialization; 3 / the provision of cheap electricity; and 4 / the emergence of a metals-based industrial complex in the Hamilton region.

Construction of the Great Western Railway, a route passing north of Lake Erie to connect the Niagara and Windsor frontiers, was completed in 1854. Allan MacNab, a major stockholder in the railway and a prominent Hamiltonian, used his influence to ensure that the route passed through Hamilton rather than above the escarpment to the south as originally planned (Roberts, 1964).

The Great Western Railway built what were the largest railway repair

TABLE 13.2
Canadian vs. U.S. steel-industry performance in the slump of 1982
(all figures in tonnes $\times 10^3$ of crude steel)

	United States		Canada	
	Output	Imports	Output	Imports
1981	109,614	18,052	14,387	2,612
1982	67,656	15,116	11,762	943

SOURCE: *The Iron and Steel Industry in 1981.* Paris: Organization for Economic Cooperation and Development 1983

yards in Canada along its tracks near the west end of Hamilton Harbour. In addition, facilities were built to re-roll British rails, which tended to crack in the cold Canadian winter. In the 1860s and 1870s, a number of other iron-based industries located around the railway yards, constituting the beginning of a local specialization in metal-working industries.

The introduction of the National Policy of the Macdonald government in 1879, which placed a 35 per cent tariff on imported hardware items, was a further stimulus to industrial growth in Hamilton. It not only protected existing Canadian manufacturers, but also gave American industrialists an incentive to locate facilities across the border in order to avoid the tariff (Kilbourn, 1960).

With the expansion of the Welland Canal in 1887, Hamilton had excellent accessibility by both rail and water to markets and resources in Canada and the United States, and was therefore an attractive location for American industrialists. For example, investors from Ohio purchased and upgraded facilities which had been used to re-roll rails and founded Ontario Rolling Mills.

Prior to 1890, 80 per cent of the pig-iron purchased by Canadian metal fabricators was imported, mostly from Scotland and the United States. In order to stimulate domestic production, tariffs were extended to cover imported pig-iron in the late 1880s. Hamilton, where coal from Appalachia and iron ore from the Lake Superior region could be assembled as cheaply as in Pittsburgh, was a natural site for iron production (Roberts, 1964).

The Hamilton Blast Furnace Company produced the first local pig-iron in 1895. This firm merged with Ontario Rolling Mills to form Hamilton Steel and Iron Company, which produced Hamilton's first steel in 1900. Thus, Hamilton entered the new century with a fully integrated iron and steel industry.

At the local level, Hamilton's municipal government supplemented the benefits of the National Policy by providing incentives to new industrial firms locating within its jurisdiction. For example, in 1890, the city finance committee approved a ten-year holiday from local property taxes for new industrial firms (Roberts, 1964). Land and capital grants were awarded to the largest firms, and the improvement of harbour facilities made possible the development of huge production facilities along the east end of the city's waterfront.

The availability of cheap electricity further enhanced the attractiveness of Hamilton to firms in the steel industry, as well as other industries. The Cataract Power, Light and Traction Company built a hydro-electricity station at DeCew Falls using water diverted from the Welland Canal in

1896. This made Hamilton the city with the cheapest electricity in all Canada at the turn of the century. Later on the availability of cheap electricity would make possible the installation in Hamilton of the first electric rolling mills in North America (Kilbourn, 1960).

The iron and steel industry does not produce goods for final consumption. Instead, it is part of a complex of industries which include manufacturers of appliances, transport equipment, and other fabricated metals industries, as well as the construction industry.

The benefits of government incentives, excellent transport facilities, and cheap power attracted firms in a broad range of industries to Hamilton during the late nineteenth and early twentieth centuries. Many firms, such as International Harvester, Westinghouse Canada, and National Steel Car, became major consumers of locally produced steel. Others, including Westinghouse which built the electric rolling equipment mentioned above, provided capital goods for the iron and steel industry. This industrial growth in Hamilton was enhanced by an efficient juxtaposition of economically linked firms.

The Emergence of Canada's Steel Capital

In 1910, Hamilton had less than a 15 per cent share of Canadian steel production. The early leader, with 40 per cent of the market, was the Dominion Iron and Steel Corporation (Dosco) of Sydney, Nova Scotia. Nova Scotia was then the most economical location for steel production because of its coal reserves and the nearby Newfoundland iron-ore reserves. Dosco specialized in the production of steel rails, which were then in great demand and protected by a high tariff. The Algoma Steel Corporation of Sault Ste Marie also concentrated on the production of steel rails (Kilbourn, 1960).

Over the subsequent decades, Canadian steel production became increasingly centred on Hamilton. In the post-war years, Hamilton's share of production exceeded 50 per cent, reaching a peak of over 70 per cent in the early 1970s (Woods Gordon and Co., 1977).

One important reason for Hamilton's ascent in the steel industry was the founding of the Steel Company of Canada (Stelco) in 1910. This was the result of a merger of iron and steel firms located in Hamilton, Toronto, and Quebec by Montreal financier Max Aitken. After the merger, most of Stelco's investments were concentrated on the Hamilton harbour facilities.

The creation of such a large corporation was a major advantage for Hamilton because of the importance of scale economies in integrated steelworks. In fact, even as late as the 1970s many foreign governments were encouraging mergers in order to improve efficiency. Stelco soon established itself as the dominant firm in the Canadian steel industry.

Hamilton's second major steelmaker, the Dominion Foundaries and Steel Corporation (Dofasco), was founded in 1912, but did not become a fully integrated steel mill until 1951. Thereafter, Dofasco quickly emerged as a major force in the industry. Between 1963 and 1975, its production increased by 120 per cent, while Stelco's production increased by 67 per cent. It is now the second-largest steel producer in Canada.

Dofasco has gained a reputation as an innovator in technology, labour relations and market strategy. In 1957, it installed the first basic oxygen furnace in North America. Today it uses this process exclusively. Dofasco is the largest non-unionized steel company in the world today. As a disincentive to unionization, it offers its workers a profit-sharing program. Although Dofasco pays competitive wages, its freedom from strikes has been a major advantage in securing a large share of the Canadian steel market. More than any other steel firm in Canada, Dofasco has benefited from the strategy of concentrating on a narrow range of products. By specializing in the production of flat rolled products, it has eliminated the need for bloom and billet mills, and achieved great scale economies in the finishing stage of production.

In the early years of the century, Hamilton was at something of a disadvantage relative to Sydney and Sault Ste Marie in terms of transport costs. Initially, some iron ore from the Lake Superior region had to be landed at Port Edward on Lake Huron and transferred to rail cars for shipment to Hamilton. In 1932, the expansion of the Welland Canal made it possible for all of Hamilton's ore to be delivered entirely by water. This effectively removed the last logistic impediment to the expansion of Hamilton's steel works. The physical expansion of the production capacity was largely accommodated by infilling part of Hamilton harbour as shown in figures 13.3 and 13.4.

Any advantage that Sydney and Sault Ste Marie had over Hamilton in terms of access to material inputs was eventually to be outweighed by Hamilton's superior access to steel-consuming markets. Unlike these competing centres, Hamilton was located in the heart of Canada's emerging industrial region – Southern Ontario's 'Golden Horseshoe.' Even before the creation of Stelco, Hamilton's iron and steel producers had important local customers. As Canada's industrial heartland shifted from Quebec to Ontario, Stelco and Dofasco found many new industrial customers. The phenomenal growth of the Toronto region in the 1960s created demands for steel products in both manufacturing and construction for which Hamilton was well situated to fill.

Figure 13.3 The Bayfront and steel mills, 1935
National Air Photo Library

Figure 13.4 The Bayfront and steel mills, 1980
National Air Photo Library

One of the most important developments for Hamilton's steel industry was the signing of the 1965 Auto Pact between the United States and Canada. By removing tariffs, while ensuring a balance of trade between the two countries, this pact made large-scale automobile production in Canada economically feasible. The huge plants built by Ford and General Motors on the shores of Lake Ontario became the Hamilton steel makers' most important customers. Stelco currently sells 22 per cent of its output to the auto makers and another 18 per cent to firms that make parts and accessories for automobiles (Hamilton *Spectator*, 18 October 1984).

Separation from these markets for steel products is responsible, at least in part, for the decline of Sydney, Nova Scotia, as a major steel centre. Dosco was taken over by the provincial government in 1969 and rearranged as a crown corporation called Sydney Steel Corporation. Today it is the smallest integrated steelworks in Canada with less than 4 per cent of the national steel-making capacity (Cordero and Searjentson, 1983). Algoma in Sault Ste Marie has overcome its isolation in part by specializing in heavy steel shapes for the construction industry. However, the end of the construction boom of the 1970s may have serious consequences for Algoma. It currently has nearly 20 per cent of Canada's steelmaking capacity (Cordero and Searjentson, 1983).

Hamilton's Current Competitive Position

Hamilton is still the largest steel-producing centre in Canada, but its dominance in the industry appears to be declining. Today, Hamilton has 60 per cent of the integrated steel-making capacity in Canada and only 47 per cent of the total steel-making capacity (including electric-furnace facilities). In light of Hamilton's dependence upon the steel industry, decline could have a serious impact on the local economy.

One cause for concern is the fact that Hamilton's steelmills use some of the oldest production technologies in the country. Much of Canada's remaining open-hearth capacity is at Stelco's huge Hilton works on Hamilton Harbour. Many finishing facilities are holdovers from the early 1960s. Perhaps the most damning illustration of the level of technology is the fact that neither Stelco nor Dofasco employs continuous-casting equipment in its Hamilton facilities. (This situation is soon to be changed.)

This technological laggardliness bodes ill not only for Hamilton's ability to maintain its share of the national steel market, but also for the sensitivity of the region to business cycles. When there is a slump in the demand for steel, it is usually the least efficient facilities that fall idle. Therefore, regions with relatively inefficient production facilities must bear the brunt of cyclical unemployment. Thus, during recent slack periods, Stelco laid off a large part of its Hamilton labour force, while workers at its newer, more efficient Lake Erie works remained on the job (Hamilton *Spectator*, 14 December 1984). Ultimately, the insecurity induced by frequent lay-offs and call-backs may be as destructive to the regional economy as permanent job loss.

Hamilton's competitive position in the industry has been eroded further by the current trend towards electric-furnace steel-making. Electric-furnace facilities operate at much smaller minimum efficient scales, and are not dependent upon the availability of iron ore and coal. Consequently these 'mini-mills' are not tied to traditional steel-producing locations. In fact, there are some reasons why mini-mills may prefer not to locate near integrated works. For one thing, integrated works consume a great deal of steel scrap. Therefore, a mini-mill located away from integrated mills can have more control over locally generated scrap supplies. Also, since many mini-mills employ non-unionized labour, they may shun towns with large and influential steel-union locals.

Hamilton currently has just slightly more than 5 per cent of the electric-furnace capacity in Canada (Cordero and Searjentson, 1983). Some industry analysts expect the proportion of steel produced by electric furnaces to continue to increase (Barnett and Schorsch, 1983). If it does, it will almost certainly cause a further decline in Hamilton's position in the industry.

Perhaps the most disturbing recent development is Stelco's decision to locate its newest integrated steelwork in Nanticoke, on the shares of Lake Erie, rather than in Hamilton. Heretofore, all of Stelco's integrated facilities have been on Hamilton harbour. Thus, the Lake Erie works represents the first serious challenge to Hamilton's dominance of the integrated steel-production sector.

The location of the Lake Erie works is roughly equivalent to Hamilton in terms of transport costs (LoSchiavo, 1984). However, the Nanticoke site has significant advantages in terms of the availability of land and the quality of labour relations.

Stelco's stated reason for choosing the Nanticoke site is the availability of a large tract of land for the construction of a 'greenfield' plant. The conventional wisdom in the steel industry is that the best way to create highly efficient facilities is to start from scratch and design a plant layout that optimizes the flow of materials rather than minimizes the requirements for space (Stelco, n.d.). There simply is not enough room for such a plant on Hamilton's harbourside. Furthermore, Stelco was able to purchase enough land at Nanticoke to allow for future expansion. It is unlikely that such a large piece of land could have been acquired at a comparable price along the western shores of Lake Ontario (LoSchiavo, 1984).

A second major advantage of the Lake Erie works for Stelco involves labour relations and technological change. Stelco's existing contract with the members of the United Steelworkers of America, Local 1005, includes a package of compensation for workers made redundant by technological change. This agreement does not apply to the new Lake Erie works (LoSchiavo, 1984). Furthermore, it is possible that more flexible work rules can be negotiated with the Nanticoke local than with the Hamilton local.

In the wake of Stelco's decision to locate its new facility outside Hamilton, there has been much speculation that the giant corporations were preparing to abandon their harbourfront facilities. Recent events indicate that this is not the case. In December 1984, Stelco announced that it would spend $400 million on three new continuous-casting operations for its Hamilton facility: a slab caster and a bloom caster by 1987, followed by a billet caster in 1988 or 1989 (Hamilton *Spectator*, 15 December 1984). Part of the funds are to be raised through what amounts to a mortgaging of the Lake Erie works. At about the same time, Dofasco announced its intention to spend $600 million over two years on continuous-casting equipment and other improvements (Mitchell, 1984).

These announcements are indeed heartening. They indicate a renewed commitment by Stelco and Dofasco to their harbourfront operators and they ensure that Hamilton will retain its leadership role in the steel industry well into the next century. However, there is no indication that the steel industry will provide any new employment opportunities in the foreseeable future. Dofasco expects its employment to stabilize close to its current level of 11,500 (Mitchell, 1985). It is unlikely that Stelco's current Hamilton employment of about 10,000 will increase. Therefore, while the steel industry will remain a crucial element in Hamilton's economy, the region's future economic growth will depend on substantial diversification.

REFERENCES

Barnett, Donald F., and Schorsch, Jarvis. 1983. *Steel: Upheaval in a Basic Industry.* Ballinger, Cambridge

Cordero, Raymond, and Searjentson, Richard. 1983. *Iron and Steel Works of the World.* Metals Bulletin Books, London

Kilbourn, William. 1960. *The Elements Combined: A History of the Steel Company of Canada*. Clarke, Irwin and Co., Toronto

LoSchiavo, Michael A. 1984. Stelco's Lake Erie Works. Unpublished BA thesis, Department of Geography, McMaster University

Mitchell, Paul. 1984. Dofasco project is linked to future. Hamilton *Spectator*, 1 December

– 1985. Auto sector boom helps propel Dofasco to a record breaking year. Hamilton *Spectator*, 5 February

Roberts, R.D. 1964. The changing pattern in distribution and composition of manufacturing activity in Hamilton between 1861 and 1921. MA thesis, Department of Geography, McMaster University

Stelco n.d. *Lake Erie Works*. Stelco, Toronto

Woods, Gordon and Co. 1977. *Hamilton-Wentworth Steel and Related Industries Substudy*. Woods, Gordon and Co., Toronto

CHAPTER 14

Energy flows and the city of Hamilton

STEPHEN C. LONERGAN

The volatility of the price and supply of energy in developed countries in the last decade has underscored the vulnerability of certain regions to these factors. Although a major energy producer, Canada exhibits the highest levels of energy consumption per capita and per dollar of gross national product in the world. Climate, the spatial distribution of the population, federally regulated prices, and Canada's industrial mix are major contributors to the high levels of consumption, but the most important example of our energy profligacy and the vulnerability of certain regions to higher energy prices may be determined by an industry-specific accounting of energy intensities (e.g., Lonergan, et al., 1985). Regions such as Southern Ontario, with a large concentration of energy-intensive industries, must be concerned with the future supply and price of this essential input to production.

Energy is not only a national problem, as expressed in a national energy policy, for example, but is a regional one as well. The consequences of changes in the price or supply of energy will be spatially variable, and the policy response will differ across regions. This essay attempts to provide an overview of the energy scene in the Hamilton-Wentworth region, with some theoretical insight and policy responses added, in an effort to promote a better understanding of the importance of energy in regional growth.

Energy consumption in industry accounts for between 40 and 50 per cent of the total energy used in developed nations. There is, however, tremendous variability in energy intensities across industries, ranging in Canada from lime manufacturing, which exhibited a total energy intensity of 280 gigajoules per dollar of output in 1982, to women's clothing manufacturers, with an intensity of less than one gigajoule per dollar of output. Industries such as cement manufacturing, pulp and paper, and iron and steel, which are important to the Canadian economy, are very energy intensive. Additionally, industries in Canada exhibit greater energy intensities than comparable industries in other countries and will undoubtedly be at a disadvantage in a future of higher energy prices and possible supply disruptions. Energy conservation or investment in less energy-intensive capital stock will need to occur in order to reduce total energy consumption. Energy-intensive capital stock in basic industries, however, is not readily convertible into energy-efficient capital. The more likely short-term changes will involve lower productivity, fewer wage concessions, or even layoffs in those sectors most adversely affected. A similar situation may exist relative to specific regions within Canada. A federal policy directive that set the price of a unit of natural gas at 67 per cent of the price of a comparable unit of fuel oil, for example, had substantial impact on the petrochemical industries in Sarnia, Ontario, and Calgary, Alberta. The Sarnia plants use fuel oil as their primary feedstock, while in Calgary natural gas is used. The potential regional impacts were evident and only with a stabilization in crude-oil prices and the eventual dissipation of this price-regulated differential has the controversy abated. It is apparent that energy prices and supply not only affect different industries but have variable consequences within an industry as well.

Although somewhat less variable across Canada than industrial mix, energy consumption in the residential, commercial, and transportation sectors is also a significant element in a region's energy economy. Variations in the first two across space tend to be the result of differences in climate, while energy used in transportation varies little between regions in Canada, any difference being primarily due to the spatial extent of the urban area and the percentage of the population using public transit.

The energy problems faced by regions in Ontario have an added dimension. Although Canada is one of only five developed countries in the world that are net exporters of energy, Ontario produces only 15 per cent of its total energy needs (Statistics Canada, 1984). This dependence on imported energy, primarily from western Canada and the United States may pose a locational disadvantage for Ontario relative to provinces with indigenous energy supplies. The need for conservation to increase the province's 'energy efficiency' is similar to the need for industries to be energy efficient; with the higher price of energy, such efficiencies may be of greater importance in the future.

Urbanized areas pose other problems. Cities in North America have expanded because of the availability of cheap subsidized energy. An abundance of high-quality energy in the form of petroleum and, more recently, electricity has led to highly centralized aggregations of people and capital; the future viability of these areas is dependent on the continued availability of cheap energy. Regions that exhibit a greater dependence on non-renewable energy may be severely stressed in times of limited availability and higher prices.

A theoretical argument also exists for concerning ourselves with a region's energy intensity. Alfred Lotka (1922), a noted biologist of the early part of this century, hypothesized that systems that maximize their throughput of energy and use this energy most efficiently will win out in competition with other systems. This implies that systems, whether social or ecological, should try to capture as much energy as possible; when energy becomes limiting, however, systems that process it most efficiently will be at a competitive advantage. The difficulties in applying this theory to regional growth and development notwithstanding, the importance of a region's energy economy is apparent; regions that are entirely dependent on an external source of high-quality, cheap energy may be severely constrained in the future as conventional energy supplies dwindle and prices rise.

The following discussion looks at energy consumption in the Hamilton metropolitan area in the context of the ideas presented above. How might Hamilton fare in the future relative to other regions in Ontario and Canada? Is the region responding to the need for more efficient uses of energy? Should Hamilton be concerned? Prior to discussing the specifics of the Hamilton-Wentworth energy economy and the responses of the region, a brief overview of the Canadian and Ontario situations is presented. Region-specific problems and responses are then outlined, followed by a discussion of the future of Hamilton-Wentworth from an energy perspective.

ENERGY SUPPLY/DEMAND IN CANADA AND ONTARIO

Canada at present consumes more energy per capita than any other country in the world. Japan, France, Britain, West Germany, and Sweden all consume less than half the energy per person than does Canada. Recent conservation efforts have reduced the total energy consumption, but subsidized prices have minimized the price-induced conservation seen in other developed countries. Figure 14.1 shows the trend in per-capita energy consumption since 1964, for both available energy (secondary energy consumption according to heat equivalents) and total energy (where the quality of the energy is taken into account). The decline in the value subsequent to 1980 is the result of a combination of conservation efforts and recession,

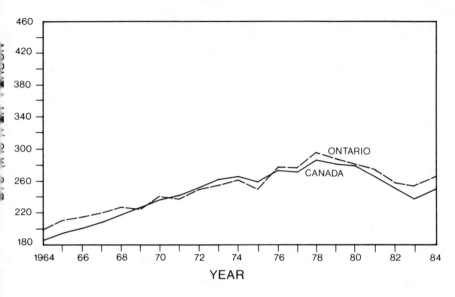

Figure 14.1 Per-capita energy consumption in Canada and Ontario, 1964-84

but how much of the decline can be attributed to each specific cause has yet to be determined.

Figure 14.2 presents energy consumption per dollar of gross national product (GNP) for 1958-82, again for available energy and total energy. The ratio of total energy to available energy provides an estimate of the conversion to higher-quality energy forms such as electricity. This ratio increased from 1:547 to 1:621 between 1958 and 1982; instead of the energy per GNP ratio exhibiting a marginal decline during this period, the total-energy curves show that Canada now uses more energy per dollar of output than it did twenty-five years ago, once the energy used in transforming from a lower to a higher energy source (e.g., coal to electricity) is taken into account. The increasing electrification of Canada and Ontario, it seems, has not resulted in greater economic output (as some have suggested may happen as economies become more electricity based), and the Canadian data suggest that quite possibly it has resulted in less.

The rich endowment of energy resources Canada has been blessed with has resulted in a relatively unique response to the higher cost of energy; rather than initially encouraging conservation efforts or allowing domestic prices to rise to world levels, energy 'mega-project' development was pursued, to ensure that domestic supply would be sufficient to satisfy

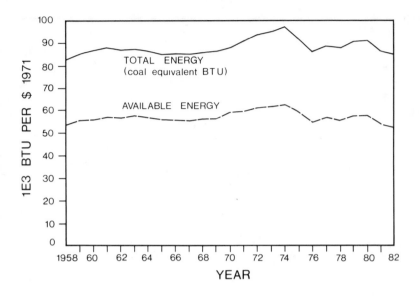

Figure 14.2 Total energy and available energy per constant dollar of gross
national product for Canada, 1958-82

future demand. Other developed countries that are large importers of
energy (as recently as 1980, Canada imported almost one-third of its crude-
oil requirements) have, however, engaged in strenuous conservation activities during the past decade, resulting in substantial declines in energy/
GNP ratios in all developed countries except Canada.

Per-capita energy consumption in Ontario has increased with the Canadian average over the past twenty years (figure 14.1), although small
declines relative to the national average are evidenced subsequent to 1978.
The provincial ratio of energy consumed to gross domestic product (GDP)
has also increased during this period, even including the marginal declines
noted after 1980. The Ontario energy/GDP ratio was 23 per cent below
the Canadian average in 1972, but this difference had decreased to roughly
10 per cent by 1982 (figure 14.3). Ontario's primary energy use (primary
energy is the energy content of the raw material) is exhibited in figure
14.4. Unlike the nation as a whole, the province of Ontario produces far
less energy than it consumes, and production now accounts for no more
than 15 per cent of total consumption (Statistics Canada, 1984). Most of
the provincial energy produced is in the form of electricity, in roughly

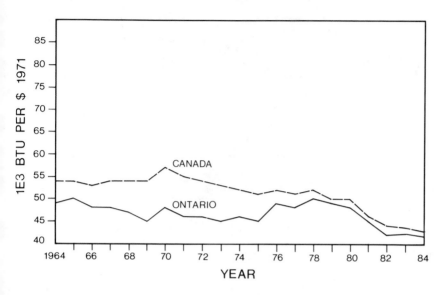

Figure 14.3 Available energy per constant dollar of output for Canada and Ontario, 1964-84

equal amounts from hydro-electric plants, nuclear power facilities, and fossil-fuel–fired generating plants. Nuclear generating plants that are scheduled to begin operations prior to 1992 are expected to increase the contribution of nuclear power to 65 per cent of the total electrical generation (Ontario Ministry of Energy, 1983).

Electricity supply to the Hamilton metropolitan area comes almost entirely from two locations – the hydro-electric facilities at Niagara Falls and one of the largest coal-fired generating plants in the world, at Nanticoke, on the north shore of Lake Erie – with small amounts coming from Lakeview (coal-fired) near Toronto and the Bruce Nuclear Power Station on Lake Huron. The remainder of the primary energy used in the province originates in the western provinces or the eastern United States. The large differential between production and consumption has motivated the province to search for alternative sources as well, and projects being promoted include the development of forest energy plantations (large stands of hybrid poplar that can be used for direct heating or to generate electricity) in eastern Ontario and co-generation facilities in the private sector. The primary emphasis in Ontario remains, however, on an expanded nuclear power program and the further 'electrification' of the province.

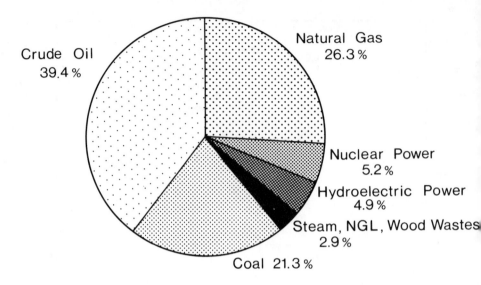

TOTAL USE: 2464.3 Petajoules

SOURCE: Statistics Canada, 57-003, 1984

Figure 14.4 Primary energy use in Ontario, 1984

REGIONAL ENERGY INTENSITIES

The manufacturing sector in developed countries typically accounts for almost 50 per cent of total energy consumption. Canadian manufacturing has consumed only 42 per cent to 43 per cent of the nation's energy supply in recent years, which is surprisingly low given the mix of energy-intensive industries in the country. The depressed economic situation in the early 1980s, however, accounts for much of the decrease in energy consumption across this sector, and the severity of the climate in Canada results in higher energy intensities in remaining sectors relative to other countries. A major reason for the high energy/GNP ratios exhibited by Canada, as mentioned previously, is the high energy intensities of the manufacturing sector. The industrial mix indicative of selected metropolitan areas within Canada, in addition, provides a corresponding variation in the energy intensities of these regions. According to the theoretical framework developed by Lotka (1922), regions that are very energy intensive may be at a competitive disadvantage in times of higher energy prices or reduced supply relative to regions with lower energy intensities. The energy intensities of selected industries in Canada are illustrated in figure 14.5, while figure 14.6 provides a corresponding comparison of the regional energy intensities of the manufacturing sector in twenty-two census metropolitan areas in Canada. Many of the industries noted in figure 14.5, such as paper and allied industries, exhibit substantial potential for energy conservation; recent data that might account for price-induced conservation subsequent to the deregulation in domestic oil prices are, unfortunately, not yet available.

ENERGY FLOWS IN HAMILTON

The Hamilton region, with its extensive iron and steel sector, exhibits an energy intensity in manufacturing exceeded by only three census metropolitan areas in the country. Does this place Hamilton at a competitive disadvantage in the future relative to other regions in Canada? Quite possibly, since capital investment in iron and steel that might incorporate energy-saving technology is occurring outside the region. This in no way implies that the energy intensity associated with iron and steel production in Canada places it at a disadvantage relative to other countries. Quite the contrary, as Lonergan, et al. (1985) have shown, iron and steel production in Canada uses less energy per dollar of output than it does in the United States (figure 14.7). The reliance on resource-extraction industries and heavy industries in general, however, does contribute to a higher energy/GNP ratio for Canada relative to other countries, and iron and steel represents one of the few sectors exhibiting an energy intensity lower than that of comparable industries in other countries. The heavy-manufacturing base of Hamilton, accordingly, poses problems for the region relative to other regions in the country. If one accepts the tenet (first developed by Lotka [1922] relative to natural systems) that regional survival is very much a function of energy throughput and energy efficiency, the economy of the Hamilton region may be more sensitive to fluctuations in energy supply and prices than other areas in Canada.

Manufacturing is the major energy-consuming sector in the Hamilton region, accounting for approximately 50 per cent of total direct energy consumption. Table 14.1 provides energy-consumption data by major-industry group for the Hamilton census metropolitan area. Although variations in climate and in the spatial development of regions may be significant

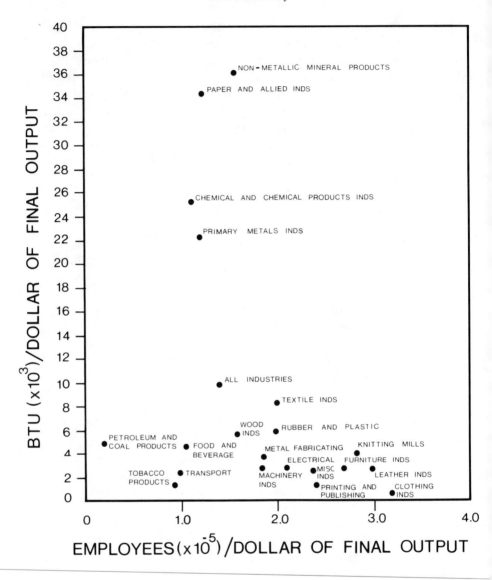

Figure 14.5 Energy/labour intensity by major industry group, Canada, 1978

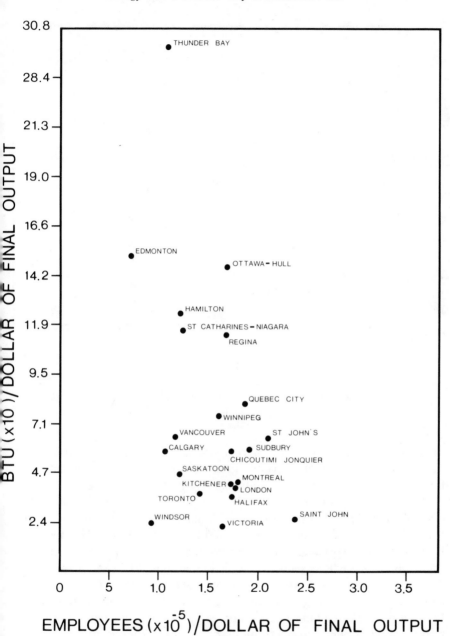

Figure 14.6 Energy/labour intensity of manufacturing in selected census metropolitan areas in Canada, 1978

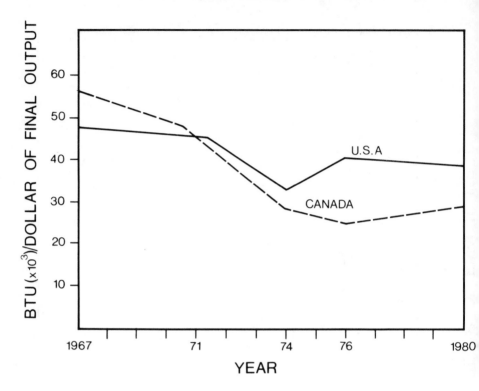

Figure 14.7 A comparison of energy intensities in the iron and steel industry in Canada and the United States, 1967-80

in determining differences in total regional energy consumption, industrial mix remains the primary variable and, accordingly, an appropriate one for speculating on regional energy consumption in the future.

The primary metals industry in Hamilton accounted for roughly 48 per cent of the value of shipments for all industries in 1980, but used more than 73 per cent of the energy consumed in manufacturing. This industry is, somewhat surprisingly, less energy intensive than primary metals in the country as a whole, as is also indicated in table 14.1 (and figure 14.5). The importance of primary metals to the region in terms of percentage of total output and energy consumption, however, is the main reason for Hamilton's high regional energy intensity, as noted in figure 14.6. The non-metallic mineral products industry, with an energy intensity twice that of primary metals, is the only other industry exhibiting an energy/output ratio greater than the regional average. This sector accounts for

TABLE 14.1
Total energy consumption and energy consumed per dollar of output by major industry,
Hamilton census metropolitan area, 1980

Major industry group	Energy consumption ($\times 10^9$ Btu)	Per cent of total	Btu $\times$ 10^3/$
01 Foods and beverages	1,903.0	2.9	3.65
04 Leather	64.4	0.1	1.72
05 Textiles	352.6	0.5	4.39
07 Clothing	37.0	0.1	0.56
08 Wood	40.1	0.1	2.81
09 Furniture and fixtures	30.5	0.1	2.65
10 Paper and allied industries	73.9	0.1	0.78
11 Printing and publishing	107.1	0.2	1.69
12 Primary metals	47,693.9	73.5	17.05
13 Metal fabricating	3,144.2	4.8	4.49
14 Machinery	1,442.9	2.2	6.13
15 Transportation equipment	580.2	0.9	4.35
16 Electrical products	1,148.1	1.8	4.21
17 Non-metallic mineral products	5,030.8	7.7	35.61
19 Chemical and chemical products	3,253.4	5.0	8.25
20 Miscellaneous manufacturing	16.2	0	0.46
Total manufacturing	64,919.4	100	

only 2.4 per cent of the total manufacturing output in the region, however, and 7.7 per cent of the energy consumed in manufacturing. The primary metals sector and the non-metallic mineral products sector together account for little more than 50 per cent of the total manufacturing output in the region, but, as noted in table 14.1, require over 81 per cent of the energy used for manufacturing and 41 per cent of the total energy consumed by the region. In all other industries, the amount of energy consumed as a percentage of total is less than the percentage contribution to total output.

THE ENERGY INTENSITY OF HAMILTON

As mentioned above, the City of Hamilton is supported primarily by a large, energy-intensive manufacturing sector. Although located in a region where the climate is mild relative to other regions in Canada, the lack of an efficient public transit system, an energy-inefficient housing stock, and the absence of a provincial energy source other than hydro-electricity pose serious problems for the future of the Hamilton energy economy. Total energy consumption in the region is not expected to grow rapidly, if at all, because of a low – and possibly negative – population growth-rate, little industrial expansion and price-induced conservation. The lack of growth in energy consumption, however, belies the energy-intensive nature of the region; only continued concern with conservation and significant investment in less energy-intensive capital stock will diminish this energy dependence.

The energy impact of an expanding service sector, additionally, is uncertain. Although it has been suggested that the service sector is, in general, less energy intensive than the manufacturing sector (particularly relative to iron and steel production, for example), this may not be true. In the process of 'regional restructuring,' moving from a manufacturing-oriented economy to a service-oriented one, we might reasonably expect the amount of energy used per dollar of income to decline. This has, however, not happened in Canada, as the energy/GNP ratio has been remarkably constant over the past twenty-five years. Between 1958 and 1982, if one accounts for the change in the percentage of electricity used, the ratio actually increased. Only with rapid escalation in the past three years has price-induced conservation resulted in lower energy/GNP ratios.

ENERGY CONSERVATION PROJECTS AND THE CITY OF HAMILTON

The primary focus of the preceding discussion has been energy consumption in manufacturing. This sector accounts for roughly 50 per cent of the energy consumed in the Hamilton region, and, owing to the variability in industrial mix across regions in Canada, provides the most appropriate base for any study of regional differences in energy consumption. The high percentage of energy consumed in manufacturing in Hamilton (substantially higher than the national average of 42 per cent) underscores the importance of this sector when examining regional energy flows. Although less variable across space, the transportation, commercial, and residential sectors are quite obviously also important energy consumers and are expected to be responsible for most of the reduction in the rate of growth of energy consumption as a result of conservation over the next two decades (National Energy Board, 1984). These sectors are particularly relevant at the regional level, because municipal and regional governments can be involved directly in decisions concerning energy-demand management. Decentralized policies can provide a major influence in reducing a region's energy intensity, both through direct subsidies and investment and through information dissemination. Hamilton has been active in the energy-management area, as evidenced by the sponsorship of the third annual Energy and the Cities Conference in 1983. Alternative mechanisms that are possible include the development of energy-efficient housing projects (see below), the consideration of energy conservation in municipal and regional plans (Regional Municipality of Hamilton-Wentworth, 1978), promotion of public transit systems, and parking regulations. Through a combination of public and private investment, the city of Hamilton has been involved in a number of noteworthy energy-conservation projects. Four of the most interesting are discussed below.

Energy-Efficient Neighbourhood Development. A 0.78-ha (1.9-ac) former school site in downtown Hamilton at present being converted to an energy-efficient 46-unit condominium development. The purpose of the development is to draw families to the urban core and illustrate that urban multi-family complexes can be extremely energy efficient and reasonably priced. The development is designed to cut in half the heating needs of homes built to present standards. The amount of exterior wall space is minimized, natural ventilation is incorporated, and exposure to solar radiation is maximized. The project, which will be completed in 1986, reduces the household energy budget to less than 50 kwh/m^2 per year (42.5 kwh/yd^2 per year).

Solid-Waste Reduction Unit (SWARU). A joint venture begun in 1981 by the Ontario Energy Corporation, the Ontario government's commercial agent for energy ventures, and Tricil Limited of Mississauga, Ontario, which operates a municipal solid-waste incineration facility in Hamilton (SWARU) is designed to generate energy from waste. Electric power will be generated by using the steam produced in garbage incineration and, subsequently, sold to Ontario Hydro for use in its grid system. The plant produces electricity from a four-megawatt generator, and results in savings of more than 3,000 kL (800,000 gal) of fuel oil equivalent per year, enough to heat 1,300 houses per year. The SWARU plant, which is the first solid-waste incineration facility in the world to generate electricity from the waste steam, is heavily subsidized by both federal and provincial governments. The facility handles one-third of the solid waste produced by Hamilton, but a controversy surrounding the amount of waste being processed has led many in the city to advocate its dissolution. The plant, however, is in the process of being upgraded and should continue as a useful experiment in the conversion of municipal garbage to a useful form of energy.

Mohawk Hospital Services. Mohawk Hospital Services, a non-profit laundry service for hospitals in the Hamilton area, has at present the largest hot-plate solar collector system in North America. Completed in September 1982, the system was designed to provide 30 per cent of the energy required to heat the 140 million litres (37 million gallons) of washing water used by the laundry each year. The heating system consists of 814 copper collectors covering over 2,000 square metres (6,500 square feet) of surface. Ethylene glycol is circulated through the collectors and a heat exchanger to heat the water. The initial year of operation resulted in a $40,000 savings over the previous year on a total cost of $1.3 million.

Retrofitting Institutional Buildings. The City of Hamilton has been a leader in pursuing energy conservation through the conversion of oil-fired units to natural-gas–based heating systems in publicly owned buildings and investing in retrofit measures to reduce total energy consumption. Municipal buildings, area libraries, and the regional court house have all undergone substantial energy conservation facelifts in recent years. Although the overall energy saving is not yet known, the city is committed to promoting conservation through a combination of energy auditing and retrofitting.

CONCLUSION

The above discussion has attempted to provide a view of Hamilton's energy economy, noting both the specific energy-conservation projects that the city is involved in as well as the importance of viewing the city's future from an overall energy perspective. Energy-consumption data in individual sectors were not presented, since they would be rough estimates only and because the uniqueness of the city lies more in the concentration of heavy, energy-intensive industry. The result is that Hamilton exhibits one of the highest energy-intensities in manufacturing overall of any region in Canada. Accepting Lotka's (1922) principle applying to regional competition implies that Hamilton should be investing heavily in mechanisms – policy or structural – to lower its energy intensity. The four projects mentioned above are examples of the city's concern for energy conservation; it is commendable that the city has taken a lead in energy planning in Canada.

REFERENCES

Calzonetti, Frank, and Solomon, Barry, eds. 1985. *Geographical Dimensions of Energy.* D. Reidel Publishing Company, Amsterdam

Lonergan, S.C. 1985. Regional development as an entropic process: a Canadian example. In Calzonetti and Solomon (1985): 393-409

Lonergan, S.C., Zucchetto, James, and Nishimura, Takao. The dynamics of industrial energy consumption: a U.S./Canada comparison. *Energy Policy*, forthcoming

Lotka, Alfred. 1922. Contribution to the energetics of evolution. *Proceedings of the National Academy of Sciences*, 8: 147-51

National Energy Board. 1984. *Canadian Energy Supply and Demand 1983-2000*. Queen's Printer, Ottawa

Ontario Ministry of Energy. 1983. *Ontario Energy Review*. Queen's Printer, Toronto

Regional Municipality of Hamilton-Wentworth. 1978. *Energy Considerations in Urban and Regional Planning*. Hamilton

Statistics Canada. 1984. *Quarterly Report on Energy Supply-Demand in Canada*. Queen's Printer, Ottawa

CHAPTER 15

Urban policy in Hamilton in the 1980s

MICHAEL WEBBER AND
RUTH FINCHER

During the twentieth century, Hamilton's natural harbour, its access to coal, iron ore, and local limestone, and its location in the heart of the Canadian manufacturing belt have ensured the city's dominance of Canada's iron and steel industry. With a population of about half a million, Hamilton ranks fourth (after Toronto, Montreal, and Vancouver) among Canada's metropolitan areas in manufacturing employment, and now is responsible for over half the total Canadian steel output.

During the 1970s Hamilton's population grew at only two-thirds of the Ontario rate and at half the Canadian rate. Even more striking, its labour force grew at less than half the provincial rate (Thompson, 1983). Between May 1981 and January 1983, Hamilton lost a quarter of its manufacturing jobs. By February 1983, the city's official unemployment rate had risen to 16.7 per cent (compared to the provincial average of 12.6 per cent); in January of that year only 54.4 per cent of Hamilton's working-age population was employed (compared to 57.4 per cent in Ontario). In the next two years there was some recovery: Hamilton's unemployment rate was 11.1 per cent in February 1985 and the labour-force participation rate was 63.5 per cent compared to 9.4 per cent and 66.3 per cent, respectively, in Ontario. By 1986, however, the economy had improved dramatically, and unemployment dropped below 7 per cent.

Hamilton is not alone in its present turbulence; it shares its recent fortunes with many old, established industrial regions in North America and Europe. Most familiar to North Americans is the shift of manufacturing firms and employment from the U.S. metropolitan areas of the 'frostbelt' to the 'sunbelt.' But evidence from Britain, France, Denmark, and other European countries indicates that manufacturing production is being spatially redistributed.

Local governments in the declining regions have been alarmed that a prolonged outflow of industrial capital would cause a long-term deterioration of their tax bases. And while tax bases decline, local governments face the rising costs of providing community service for those people thrown out of or unable to find work – with little help from senior levels. Local business people are afraid that economic decline will reduce their bases of profitability.

In this essay, we describe and evaluate some local-policy responses in Hamilton to the 'hard times' that accompany such an industrial restructuring. After describing the characteristics of Hamilton's economy and labour force, we take as case studies two planning issues recently discussed by Hamilton's local government bodies: development of the South Mountain and the Red Hill Creek expressway. These planning policies were designed to facilitate economic development in the Hamilton metropolitan area and to make new areas accessible for residential, industrial, and commercial growth. We conclude with some analysis of the limits to the choices presented for the future development of the city.

THE ECONOMY OF HAMILTON

The economy of the Hamilton metropolitan area is dominated by manufacturing industry; furthermore, within the manufacturing sector, its work force is is inordinately concentrated within a single sector and within productive rather than administrative jobs. These types of jobs have been increasing in number relatively slowly during the 1970s and have proved to be susceptible to the effects of the depression that has characterized the early 1980s. This section briefly describes these characteristics of the Hamilton economy as they provide the background to current local-government activities.

The occupational structure of the regional municipality is described by the 1971 and 1981 censuses (Statistics Canada: *Census of Canada, 1971, 1981. Occupations*). Table 15.1 shows that, in 1971, the region had pro-

TABLE 15.1
Occupational structure of the Hamilton-Wentworth regional municipality, 1971 and 1981
(percentages)

| | 1971 | | 1981 | |
	Hamilton	Ontario	Hamilton	Ontario
White collar	15.7	20.2	22.4	23.8
Sales and service	39.5	43.4	40.0	46.0
Blue collar	44.6	36.4	34.6	30.1

portionately fewer 'white-collar' (managerial, administrative, scientific, teaching, or health occupations) workers than the province, fewer sales and service workers, and considerably more 'blue-collar' (processing, fabricating, machining, assembling, construction, transport, and materials handling) workers. By 1981, the regional municipality had become more like the province.

Several other features of Hamilton's industrial-employment structure are revealing. Hamilton's manufacturing employees are 80 per cent production workers, compared to 72 per cent for the province as a whole. A far smaller proportion of manufacturing employees are female in Hamilton than in Ontario as a whole (only 14.6 per cent of the production workers and 25.1 per cent of the administrative workers, compared to 23.3 per cent and 29.9 per cent for the province).

Hamilton's workers are, however, or perhaps as a result, on average relatively well paid: production workers earn nearly 90 cents per hour more than the average provincial worker. The industries in which Hamilton workers earn more than their provincial counterparts are the printing and publishing, primary-metals, metal-fabricating, and chemical industries.

Hamilton's industrial structure changed slowly over the 1970s and became more specialized. In the years 1971-9, the overall productive work force grew by 12.4 per cent, compared to a provincial rate of 17.2 per cent. The primary-metals and chemicals industries grew relatively rapidly in the city, having employment increases of more than 20 per cent. In addition, 8 per cent of the administrative jobs vanished, compared to an increase of 7 per cent in the province. Hamilton's characteristic reliance upon blue-collar employment increased rather than diminished during the 1970s.

The relatively slow growth of employment, and in particular the poor performance of the non-primary metal industries, prompted the initial concerns of the regional municipality with industrial employment. Yet, slow growth in the 1970s became a catastrophic decline in the 1980s. Manufacturing employment reached a peak in May 1981 in Hamilton. Since then, employment in both the city and the province has fallen: by January 1983, Hamilton had lost nearly 18,000 manufacturing jobs, almost a quarter of its total (while Ontario had lost 16.3 per cent of its manufacturing employment). Table 15.2 reveals the changes that took place in the major plants in the Bayfront Industrial Area (see figure 15.1 for its location). Even in the absence of major bankruptcies or capital flight, and even if output levels recover, technical changes in these steel-producing and -using industries promise to reduce employment even further.

TABLE 15.2
Employment forecasts of firms in Bayfront Industrial Area

Firm/Plant	1981	Spring 1985	Future
Stelco			
Hilton	14,736[a]	10,108[a]	7,500‡
Parkdale	600[b]	525[d]	525*
Frost	300[b]	160[d]	160*
Canada Drawn	140[b]	140[d]	140*
Canada	800[b]	100[d]	0†
Dofasco	11,300[b]	11,400[d]	9,500§
International	1,830[a]	730[a]	800*
Harvester			
Firestone	2,070[c]	1,680[a]	1,680*
National	1,700[b]	580[a]	600*
Steel Car			
TOTAL	33,476	25,423	20,905

SOURCES: (a) personal communication with the press or personnel managers; (b) *Scott's Industrial Directory*, 13th edition; (c) *Scott's Industrial Directory*, 14th edition, and *Hamilton-Wentworth Business Directory*; (d) *Scott's Industrial Directory*, 15th edition
* Assumed to continue at present levels
† Stelco to close plant
‡ Assuming output rises to 3 million tons at 400 tons/employee (below best Japanese current practice)
§ 4.2 million tons (current capacity) at 400 tons/employee, less 1000 administrative staff not employed in Bayfront

LOCAL GOVERNMENT PLANNING RESPONSES

There are two levels of local government within the Hamilton region. The Regional Municipality of Hamilton-Wentworth consists of a federation of municipalities: the City of Hamilton and the municipalities of Ancaster, Dundas, Flamborough, Glanbrook, and Stoney Creek. Both the region and the city have planning responsibilities, and the examples below examine decisions of both the city and the region.

People, corporations, and local governments have been forced to react to the slow economic growth of Hamilton during the 1970s and to its decline during the 1980s. Essentially, within Hamilton this reaction has taken the form of the a strong pro-development coalition of business and local government. This coalition has promoted infrastructural investments by the local governments that have the ostensible aim of attracting business investment to the city. In the short term, however, the beneficiaries may be the promoters. Two such investments in particular are the East-West / North-South freeway system (see discussion below) which is a

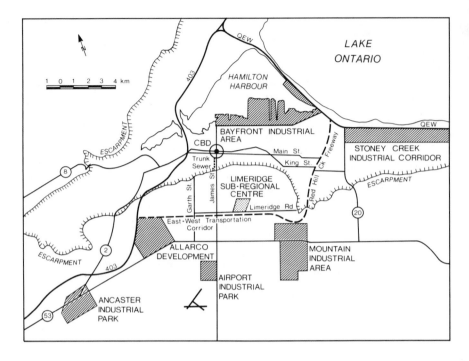

Figure 15.1 Location of developments in Hamilton

project of the regional municipality, and the construction of a trunk sewer to the South Mountain area (see discussion below), which is a project of the City of Hamilton.

Many groups within the city have an interest in promoting such infrastructural investments. Most directly, the city's construction industry stands to build the freeway and the sewer; together the projected cost of the two projects is about $210 million. Furthermore, if the projects do actually have the effect of boosting employment in the region, the construction industry and local developers can expect to build new subdivisions and work places. The support given to the freeway proposal by the Hamilton Construction Association in evidence to the Ontario Municipal Board shows this interest. But the local governments themselves are not acting solely in response to pressures from the development and construction industries, for they have their own needs to meet, too.

The financial health of many North American local governments has been threatened by the decline in manufacturing activity and labour force in their jurisdictions. Largely because of the decline of local property taxes in the United States, says Badcock (1984), 'councils are preoccupied with attracting private investment to fortify the local tax base and increase revenues (or, conversely, with halting private disinvestment which depletes per capita revenues).' Similar forces act in Canada.

The competition for private-sector investment dollars has become obvious in the United States in the 1970s and 1980s. Local governments began public advertising campaigns, funded by taxpayers' dollars, to indicate how their region is a better location for new business than are neighbouring areas: Goodman (1979: 12) notes that, since the 1950s, U.S. local and state governments have come to offer a vast array of incentives for private development. These include low-interest bonds, tax exemptions on new equipment, government financing for plant expansions, tax abatements and waivers, and government provision of physical infrastructure required by private developments (e.g., parking lots, sewers, new roads). The only problem with these incentive strategies is that the cost of providing them accrues to local taxpayers – only if the incentives are followed by enough private development to offset their cost do taxpayers benefit. Profit from the new investments accrues to the investors, however, not to the taxpayers. And if the incentive strategies are unsuccessful, then the local tax base is even further depleted than it was when the incentive strategy was first implemented.

Is the same true of Hamilton? As we have seen, Hamilton's industrial base has recently declined. Has this had the effect of turning Hamilton's local government agencies into 'entrepreneurs' of the type described by Goodman for the United States? Certainly this was one of the recommendations of the 'Action Plan for Economic Growth' produced for the region by Currie, Coopers and Lybrand (1978); the region now spends nearly a million dollars per year on its promotional agency, the Economic Development department. The way in which Ontario's municipalities, including Hamilton, scrambled to get the Toyota plant (which located in Cambridge) indicates just how concerned they are for new developments. The local property tax is certainly the major source of revenue for municipalities in Ontario and has been since 1973 (Social Planning Council of Metropolitan Toronto, 1979: 25), and Ontario has a higher-than-average (for Canada) dependence on local taxes. Of its combined provincial and municipal revenue, 17.4 per cent was raised through the property tax between 1976 and 1977, for example, while the average for the other nine provinces was 12.8 per cent. Despite increases in grants from the provincial government through the 1960s to 1973, this proportion has not since declined.

Figures for all Ontario municipalities show that the proportion of their revenues made up by property taxes has, through the 1980s, been slowly increasing to match the figures for the late 1960s: by 1984 this proportion was 41 per cent. Though statistics for individual municipalities have not been published since 1979, Hamilton's level of dependence on property taxes for municipal revenues was very similar to that of Ontario municipalities as a whole for the years 1969 to 1978 (Statistics Canada: *Local Government Finance*, Catalogue 68-204 and 68-203, Annual).

All of Ontario's municipalities, like Hamilton, face fiscal difficulties in paying for education, public works, and a share of transport, welfare, and health costs, to the degree that their property tax base is depleted and not replenished by transfers from other levels of government. Though the degree of reliance on the property tax is not as great in Ontario as in the U.S. jurisdictions discussed by Badcock (1984), nevertheless, Ontario municipalities need those revenues. It is within this context of Hamilton's need for a substantial local tax base to support its population in a time of economic decline that we try to explain two controversial planning strategies in the Hamilton region.

The Red Hill Creek Expressway

For over a decade, discussions continued between the city, the region, and the provincial transport ministry over the question of constructing and locating an east-west roadway system parallel to the Niagara escarpment and north-south across the escarpment into the east end of Hamilton. Figure 15.1 shows the location of the proposed roadway, with respect to the sites of heavy industry which dominate the local economy, and the existing transport network. The projected locations of industrial growth which underlie the regional government's case that the freeway is necessary to the economic development of Hamilton are also shown. Numerous planning studies have investigated the need for the road and its place in the projected economic growth of the Hamilton region. But considerable opposition, both to the project and to certain of its characteristics, emerged within the local community: in particular, opposition has focused on environmental damage to the Red Hill Creek Valley from the north-south leg of the road, and on the validity of population and employment projections used to justify the road. (The Hamilton *Spectator*, in various issues through 1985, records much of this argument.)

The Regional Municipality of Hamilton-Wentworth (the proponent of the freeway) defended the route selected and included it in the region's official plan against a request by the Save the Valley Committee (an opposition coalition) to amend the official plan to delete reference to the proposed facility. The hearing before the Ontario Municipal Board, an arbitration body that adjudicates proposed amendments to regional official plans, led to a 2:1 majority decision for the roads. Essentially, two accepted on the basis of economic need for the road, but one objected on the basis environmental factors. Here we examine those economic arguments.

The regional government viewed the roadway as vital to the future economic growth of the region. This view reflects advice given by consultants, who have described the region's economic plight and the problems associated with its continued dependence on the steel industry, and have argued that economic diversification and revitalization will not proceed without this particular new highway system. The firm Currie, Coopers and Lybrand has been commissioned twice by the region, initially to develop and then to review an economic strategy. In 1978, the firm identified two constraints to economic development: a shortage of serviced industrial land with good highway access, and poor marketing of the region's strengths (Currie, Coopers and Lybrand Ltd., 1984: ii). The firm's 1984 review focused on the relationship between the proposed expressway and the region's economic-development possibilities. It concluded that the one remaining problem was that the Mountain Industrial Area (at 607 ha [1,500 ac] constituting nearly one-third of the region's vacant industrial land) still had inadequate highway access, and that the proposed roadway would resolve this problem. But they said more:

Moreover, we believe the E-W/N-S facility will act as a catalyst for commercial, industrial and residential development in the Region. It will link the Region's industrial/business/commercial areas into a viable, efficient network, provide for efficient movement of trucks, enhance regional labour market accessibility and serve new residential developments on the Mountain. More specifically, the facility will enhance development in the Mountain Industrial Area, the Allarco Business Park, the Lime Ridge Sub-Regional Centre, the Airport and Ancaster Industrial Parks and the Stoney Creek Industrial Corridor, and reinforce 8v the regional role of the expanded Hamilton Civic Airport. [See figure 15.1 for locations of the sites.]

We estimate that 9,340 to 12,425 permanent jobs will be accommodated in new commercial/industrial development stimulated by the E-W/N-S facility and that the estimated value of this 'net' new construction, in 1984 dollars, is between $496 and $707 million. Improved access to and across the Mountain will also stimulate new residential development estimated at approximately 300 new houses per year or 3,000 new houses over a ten-year period with a construction value of $225 million.

The proposed facility is, therefore, a necessary public investment to make and keep the Region's infrastructure competitive and is a strategic link in transforming the Regional economy. (Currie, Coopers and Lybrand, 1984: iii-iv)

The regional government's second rationale for the East-West / North-South transport corridor arises from predictions of internal travel demand, specifically demand for commuter travel. The argument is as follows.

First, the region's population is expected to grow to 445,000 by 2001; in 1982 it was 410,500. This modest population change is expected to create significant growth in the urban area south of the escarpment bounded by Limeridge Road, Highway 403, Highway 53, and Highway 20 (see figure 15.1). Yet, 'the expansion of employment is expected to remain concentrated principally within the present major employment centres located north of the Escarpment. In excess of 55 per cent of the total employment within the Region is focussed [sic] on the Hamilton Central Business District, the Bay Front industrial complex, McMaster University, and the Stoney Creek Industrial Park. It is expected these major activity centres will be reinforced. The core area will continue to be the major activity centre in the Region' (Regional Municipality of Hamilton-Wentworth, 1982: I: 2-8).

Second, following the forecast of employment growth north of the escarpment and residential growth south of it, shown in table 15.2, the travel demand crossing the escarpment is forecast to increase 65 per cent by 2001. The present roadway network across the escarpment, both for trucks and non-commercial vehicles, has been assessed as inadequate. Table 15.3 shows the anticipated travel demands which account for this assessment. Despite the fact that core area employment is expected to remain significant, and that the major industries there provide 'extensive work staggering through the 3-shift program,' little potential to increase work staggering and thus to spread work trips through the day is seen to exist in the industrial sector (Regional Municipality of Hamilton-Wentworth, 1982: I: 3-28).

Third, the region's population and employment forecasts are based on the assumption that current trends to higher labour-force participation rates will continue. Work-trip rates per person in the peak hour have been assumed to increase at the same rate, and so auto ownership is expected to remain at or above present levels though a modest increase in transit use is forecast (Regional Municipality of Hamilton-Wentworth, 1982: I: 3-29).

All told, then, the regional government has rationalized the construction of the freeway on the grounds of the need to link residents south of

TABLE 15.3

Summary of escarpment travel demand: past, current, and projected volumes, pm peak hour, peak direction

Year	Population Mountain growth area (regional population)	Households Mountain growth area	Person trips	Transit person trips (%)	Vehicle person trips	Vehicle volume per hour
1971	117,000 (391,800)	34,400	14,500	1,900 (13.1%)	12,600	9,600
1976	124,500 (409,800)	38,600	16,100	2,100 (13.0%)	14,000	10,800
1980	129,300 (410,500)	42,500	19,400	2,600 (13.4%)	16,800	12,800
2001	177,000 (445,000)	66,500	32,200	5,400 (16.8%)	26,800	19,400
Designated development capability	216,700 (550,000)	84,300	38,700	9,300 (24.0%)	29,740	21,300

SOURCE: The Regional Muncipality of Hamilton-Wentworth, *Environmental Assessment Submission*, vol. 1 (1982): table 3.4

the escarpment with jobs in the core area north of it: the opposite of the consultants' view of the need to provide highway access for the industrial parks to the south. How valid is the proposed roadway system as a solution to the region's problem as posed? We comment both on the assessment of the region's plight made by the regional government and its consultants and on the adequacy of the roadway as a means of alleviating the problem.

The region's assessment of the spatial patterns of Hamilton's probable future growth has many flaws. Since this assessment forms the basis for the need for the freeway and its proposed route, then the questioning of the assessment automatically calls into question the validity of the freeway 'solution.' The problems in the region's position are as follows.

First, the business parks located south of the escarpment (see figure 15.1) are projected to provide many jobs, and to contain industrial and other commercial concerns. Retailing is also being encouraged there (see below). It therefore seems questionable to assume that retailing jobs will increase in the core area north of the escarpment, and certainly the forecast increase in central business district (CBD) employment of 60 per cent by 2001 is overoptimistic. At the the same time, the population of the city that lives north of the escarpment is projected to decline by 24,000 (about 12 per cent) by 2001 (see table 15.4).

Second, we know that bayfront industrial jobs cannot be assumed to increase as anticipated (see table 15.2). The five firms listed in table 15.2 now employ some 80 per cent of the Bayfront work force, so their forecasts for 2001 imply a labour force in the Bayfront Industrial Area of about 25,900. Traffic from residential areas south of the escarpment to manu-facturing jobs in the bayfront therefore must be expected to decline in the next couple of decades. And if new industrial and commercial growth is to be anticipated in the business parks south of the escarpment, then the flow of traffic to work places in the north and core areas of the city may be expected to decline further.

Because of the anticipated development of new sites for commercial and industrial production, then, the two major traffic flows across the escarpment, to the central business district and the bayfront, must be expected to decline in the planning period under discussion. If the ex-pressway does lead to the creation of new jobs and business south of the escarpment, as the region's consultants predict, then it is precisely this that will reduce use of the proposed escarpment-crossing road, because people will not need to drive north to work or shop.

Third, we question whether it is reasonable to assume large residential growth south of the escarpment. The region's population growth is ex-pected to be modest, as table 15.3 shows. Should we expect to see a decline in population in the 'lower city' of over 20,000, and an increase on 'the Mountain' of about 20,000? This would involve relocation of over 12 per cent of the residents who currently live north of the escarpment. We question whether a population-redistribution of this magnitude is likely. Even if such a migration does happen, the people who move into new homes south of the mountain are more likely to be those employed in the new commercial sites south of the escarpment than those who work in the central core and bayfront.

We can summarize this argument as follows. The region forecasts that the number of employees living on the mountain will increase by nearly 80 per cent between 1980 and 2001 (comprising a 39 per cent increase in population and an increase in the proportion of the population in work of 30 per cent). This is despite a forecast population increase of 7 per cent. Equally, the region forecasts that the number of jobs located north of the escarpment will rise by 40 per cent. Together these imply that the demand

TABLE 15.4

Projected population and employment

Municipality	Population				Employment		
	1980	2001	Designated Development capability		1971	2001	Designated Development capability
Ancaster	14,400	31,500	38,000		2,000	9,400	12,000
Dundas	19,500	20,000	35,000		4,600	5,500	6,600
Flamborough	24,200	29,500	30,000		3,400	5,600	6,500
Glanbrook	9,700	9,000	14,600		1,500	3,300	7,000
Hamilton							
Lower City	195,000	171,200	216,800	Core Area	30,100	49,900	67,800
Mountain	110,900	130,800	150,000	Bayfront	46,800	48,900	54,600
Subtotal	306,800	302,000	367,000	McMaster	5,800	7,300	10,200
				Sub-regional	200	3,300	4,400
				Mountain Industrial Park	300	4,900	7,400
				Other	51,700	65,400	84,500
				Subtotal	134,900	179,700	228,900
Stoney Creek							
Lower Town	28,000	34,700	42,900	Industrial Park	2,000	12,100	12,900
Mountain	7,800	18,300	32,500	Other	4,800	12,300	16,100
Subtotal	35,900	52,600	75,400	Subtotal	6,800	24,300	29,000
TOTAL	410,500	445,000	550,000		153,200	227,800	290,000

SOURCE: Regional Planning Department; The Regional Municipality of Hamilton-Wentworth, *Environmental Assessment Submission*, vol. 1 (1982)

for peak-period travel across the escarpment will rise by 66 per cent by 2001. In contrast, we expect the population on the mountain to rise by no more than 30 per cent and a constant proportion of the population in work; the number of jobs located north of the escarpment will fall unless core employment growth can outstrip bayfront decline. Furthermore, in 1971, there were 48,800 jobs in the Bayfront and Stoney Creek industrial areas; the forecast for 2001 is 38,000: yet it is precisely these employment locations in the eastern part of the city that would be served by the Red Hill Creek Valley expressway.

None of the consultants' reports attempted to measure the effect of the roadway on new industrial development, preferring to rely instead on comparison between Hamilton's slow growth and the rapid growth of Toronto's suburban municipalities (such as Mississauga and Oakville), of Hamilton's own suburban municipality (Burlington), and of some small cities in the area (Waterloo and Guelph). However, some progress can be made towards this measurement. The first part of this task is to estimate the effect of the new road system on the transport costs of manufacturers in the region. Data on the destinations of trucks are available from a 1972 origin-destination survey (reported in the Regional Municipality of Hamilton-Wentworth's *Environmental Assessment Report* of 1982, 3-13). Data on internal travel times with and without the facility are given in Miller, O'Dell, and Paul (1984). It is assumed, in addition, that trips to Toronto and Niagara require sixty minutes in addition to internal travel and that trips in the 'Other' category require three hours in addition to internal travel. These assumptions imply that for the average manufacturing firm in the region the transport facility will reduce truck travelling times by 0.95 per cent. If all transport is by truck and if truck running costs are proportional to time, then this implies that the facility will reduce transport costs for the region's firms by 0.95 per cent. (Depending on the industry, between 10 per cent and 60 per cent of shipments are made by rail, so this calculation greatly overestimates the effect of the facility on transport costs.)

The effect of such a reduction in transport costs has been measured by the Harris and Hopkins (1972) regional econometric forecasting model. (No equivalent model seems to be available for Canada.) This model regresses the annual change in output in each industry in each county of the United States on such variables as transport costs, agglomeration economies, wage levels, and labour availability. The estimates are cross-sectional. We have proceeded as follows. First, industries analysed by Harris and Hopkins in the three-digit U.S. Standard Industrial Classification are aggregated to conform to the Canadian two-digit industrial classification. The transport-cost coefficients (on inputs as well as outputs) for all three-digit industries are summed to estimate the effect on the output of each two-digit industry of a change in transport costs. (Some two-digit industries, not represented in Hamilton, are excluded.) Column (1) of table 15.5 contains these coefficients.

The remaining columns of table 15.5 reflect computations from these basic data. Column (2) measures the cost of transport as a percentage of the value of output. Since the regression coefficients are measured in units of dollars of output change per cent of transport cost, the product of columns (1), (2), and the 0.95 per cent transport cost improvement yields column (3), the estimate of the output increase in each two-digit industry in a typical U.S. county caused by a 0.95 per cent reduction in transport costs. Multiplying this by the proportion of output that is spent on wages (column [4]) yields column (5), the expected increase in wages caused by the reduction in transport costs, in 1966 U.S. dollars. By the end of the 1970s, each industry in Canada had an output between 10 per cent and 20 per cent of the value of output of its U.S. counterpart in 1966 (see column [6]), so the expected increase in wages caused by the transport-cost reduction is given by column (7).

The idea underlying this calculation is that a reduction in transport costs in one county will allow that county to increase its output level at the expense of other counties. By scaling these output levels to reflect the size of the Canadian economy, it is estimated that the freeway construction might lead to an increase in wages of about $100,000 (1979 Canadian dollars), or an increase of about ten jobs per year, in contrast to the Currie, Coopers and Lybrand (1984) prediction of 9,000-12,000 permanent jobs. No reasonable multiplier would give rise to the estimate of Currie, Coopers and Lybrand. (No attempt has been made to measure the effect on jobs and income of building the freeway itself: any expenditure of $195 million on construction would have a similar effect.)

Upper James Street–South Mountain Development
Each municipality in Ontario is required by law to produce an official plan that is approved by the provincial government. This plan consists of a statement of the expected developments of population and employment, the location of these developments, and the policies needed to provide infrastructure for them. These plans are for 25-year periods, which in Hamilton's case run until 2001.

The official plan of the Regional Municipality of Hamilton-Wentworth

TABLE 15.5
Calculation of effects of transport facility

Industry	B coefficient (1)	Transport (2)	Output increase (3)	Wages (4)	Wage increase × 1000 (5)	Canada (6)	Wage increase (7)
Food and beverages	4,157.94	2.053	81.080	15.30	12.41	20.78	2,579
Rubber and plastics	1,022.31	1.153	11.194	26.89	3.10	20.60	639
Leather	79,685.63	1.109	839.887	32.38	271.99	14.75	40,121
Textiles	854.56	1.224	9.938	25.72	2.56	11.93	305
Knitting clothing	10,441.87	0.591	58.579	32.04	18.77	12.74	2,391
Furniture	301.05	1.435	4.103	32.91	1.35	22.86	309
Printing/publishing	422.30	1.627	6.529	36.81	2.40	17.08	410
Primary metal	17,737.68	2.874	484.239	19.67	95.24	20.70	19,717
Metal fabricating	1,114.00	1.650	17.468	28.11	4.91	26.65	1,309
Machinery	11,776.75	1.311	146.621	28.44	41.97	12.30	5,128
Transport equipment	6,489.91	1.135	69.961	17.80	12.45	19.04	2,371
Electric products	90,906.91	1.038	896.342	29.97	268.59	10.03	26,930
Non-metal minerals	11,058.50	3.144	31.617	26.14	8.27	24.05	1,987
Chemicals	2,912.51	2.224	61.541	19.62	12.08	15.17	1,832
Miscellaneous	8,181.75	0.995	77.318	29.76	23.08	15.11	3,487
TOTAL							109,515

NOTES:
(1) B coefficients from Harris and Hopkins (1972)
(2) Transport costs as a percentage of the value of output, from Canadian input-output tables
(3) Increase in output due to 0.95 per cent transport cost saving
(4) Wages as percentage of output (same source as 2)
(5) Wage increase due to transport cost saving (product of 3 and 4)
(6) Output of this industry in Canada now as a percentage of the output of this industry in the United States at time of Harris's study
(7) Wage increase in this industry (product of 5 and 6), in dollars

was developed in the mid-1970s. It assumed that the population of the region would rise to 550,000 by 2001. Since the area of the city below the escarpment is now virtually fully developed, such an increase in population (more than 30 per cent above the actual 1982 population) would have to be accommodated in the South Mountain area. Hence part of the plan called for the construction of a trunk sewer in the third and last stage of the plan and for the detailed planning of neighbourhoods and shopping facilities in this area. By 1981, the region's Planning and Development department had revised the population forecasts in the light of the region's economic performance to 445,500 people (only 7 per cent above the 1982 level).

However, instead of postponing consideration of the trunk-sewer development, the pro-development coalition in the region argued that services should be provided to the Upper James–South Mountain area even earlier than called for in the plan. This argument was given some force by the decision of the federal government to upgrade Hamilton's airport. The airport was small, providing service to the Niagara Peninsula, Ottawa, and Montreal; one runway was lengthened to allow the diversion of some charter traffic from Toronto's International airport to Hamilton. To examine the proposal to provide services to the Upper James–South Mountain area before the planned time, the City Council commissioned consultants to study the demand for commercial facilities in the Upper James–South Mountain area. The study was completed in spring 1984.

This study by Miller, O'Dell, and Paul examined a variety of population and employment forecasts to derive the demand for commercial facilities in each of the zones into which the region was divided. Demand was computed from a spatial-interaction model that incorporated multi-stop, multi-purpose trips; with the solution of the model, the size of each commercial centre was made proportional to the total sales of that centre, following a suggestion of Harris and Wilson (1978). The model is described in O'Kelly (1981a, 1981b), and it was calibrated on data collected in 1978, when 704 households in the region kept a travel diary for two weeks each.

The results of the modelling exercise are sensitive to assumptions about both the population of the region and the needs of each person for commercial space. Since increases in the city's population will be largely concentrated on the urban fringe of the South Mountain, any variation in the assumptions about the size of the population in 2001 will severely affect the expected demand for commercial facilities there. Equally, the expected total demand for space will depend on the amount spent by each person and the total sales per square metre that must be made by facilities. However, under the 'most likely' population projection of 445,500 and the median 2 projection of space needs per person, the study concluded that only 3,600m² (11,800 ft²) of commercial space would be needed in the Upper James–South mountain area in addition to the commercial space already developed on the mountain. This is the size of a neighbourhood shopping centre, which alone would not justify the construction of a trunk sewer.

The consultant's report also assessed the likely effect of upgrading the region's airport on the demand for development in the Upper James–South Mountain area. Essentially, this involved examining the development around other regional airports in Canada. Apart from jobs in the airport itself (running the airport, maintenance, catering), small airports no more than 15-20 minutes from downtown attract little in the way of hotel, restaurant and industrial development.

The report of Miller, O'Dell, and Paul (1984) thus concluded that there was little ground to anticipate much demand for development in the Upper James–South Mountain area within the horizon of the region's official plan. Despite this conclusion, the city council decided to go ahead with the plan to build the sewer and to open the area for development. In order to avoid having to justify such a change in the official plan to the Ontario Municipal Board, the city council decided to pay the $15 million cost of the storm and sanitary sewers out of Hamilton's current-account budget rather than from its capital-account budget. In order to provide these services to the Upper James–South Mountain area, the city had to post-pone several other sewer projects, including two that would have reduced sediment pollution of the harbour (Hamilton *Spectator*, 12 April 1984).

CONCLUSIONS

Our analyses present two cases in which local governments appear to be pressing on with the expansion of the region's transport and service in-frastructure. On some criteria, both the Red Hill Creek expressway and the trunk-sewer extension seem unjustified. There appear to be major flaws in the reasons given by local-government proponents of the projects. Why then, despite evidence that the projects will not have the effects claimed for them, are local governments persisting? It seems that several factors contribute to the focus of contemporary urban policy on extending the region's physical infrastructure in order to attract new investment.

The fact that decline in the local economy will continue unless the present trends are reversed means that new sources of investment are needed if the Hamilton region is to return to the economic prominence it enjoyed in the 1960s when its heavy industrial base was thriving. Not only do certain groups of local business people have an interest in commercial or residential expansion (real estate and construction interests, and local merchants, for example), but local governments themselves have an interest in the more secure local tax base which would accompany new developments. Thus, if expansion in the service infrastructure of the community can be linked to future economic growth, then local business people, government, and presumably the many working people not directly disadvantaged by the expansion can be expected to favour it.

Local governments must therefore make some attempt to improve Hamilton's economic fortunes. Even though these infrastructural investments may not provide all the benefits claimed for them, nevertheless they are practical, existing proposals. Alternatives to such major investments in infrastructure rarely emerge in the local political arena. As one author has argued in an analysis of local government spending on physical infrastructure to attract private developers to downtown Winnipeg: 'There is little political defence against the idea that building new and large complexes is progress and that to stand in the way of such endeavours is negative and destructive thinking about the city ... the concept of a major private development, strongly supported with public funds, was consistent with the political and economic elite's notion of what should be done with the city. No counter idea emerged during arguments about proper strategies for urban development. The latter in fact could be interpreted as the precondition that made it easier for the deal to be struck' (Walker, 1978: 156). In Hamilton, like Winnipeg, no concrete proposals for ways to attract new development have emerged to challenge the notion that new development may follow government expenditure on new urban infrastructure. So any challenge appears somehow negative in the 'boosterist' atmosphere of local political discussion undertaken by the local business-government coalition. Even expert evidence against a development proposal can have no effect in the absence of alternatives equally acceptable to the local boosters and their interests. The possibility that Hamilton might become a smaller urban centre, with a less specialized (but less influential) economic base, is not acceptable.

Furthermore, it is not easy for urban government agencies to make decisions which are fiscally wise. It is hard for their staff to predict the multiplier effects and revenue impacts of public expenditures when they lack control over private-sector economic decisions (Friedland, Piven, and

Alford, 1984: 282). Equally, government agencies often rely for analyses and predictions on the reports of planning consultants, who have a vested interest in recommending public planning actions that will create more work for themselves! Government agencies and locally elected officials must also be seen to be doing something, and in the absence of obvious alternatives (given acknowledged economic trends) the construction of new infrastructure is at least an attempt to help the situation. And, once planning and discussion of a large public-program expenditure has been underway for a long time (as that for the Red Hill Creek expressway certainly has been), those groups expecting to benefit from the construction are unlikely to be pleased with a local government decision to scrap the project or modify it greatly. A certain reluctance on the part of local government to abandon their long-term plans, then, might be anticipated.

REFERENCES

Badcock, B. 1984. *Unfairly Structured Cities*. Basil Blackwell, Oxford

Currie, Coopers and Lybrand Ltd. 1978. *The Regional Municipality of Hamilton-Wentworth: An Action Plan for Economic Growth*. Report to the Regional Municipality of Hamilton-Wentworth, Hamilton

– 1984. *Region of Hamilton-Wentworth: Regional Economic Development and the Proposed East-West and North-South Transportation Corridor*. Report to the Regional Municipality of Hamilton-Wentworth, Hamilton

Friedland, R., Piven, F.F., and Alford, R.R. Political conflict, urban structure and the fiscal crisis. In Tabb and Sawers (1985): 273-97

Goodman, R. 1979. *The Last Entrepreneurs*. Simon and Schuster, New York

Harris, B., and Wilson, A.G. 1978. Equilibrium values and dynamics of attractiveness terms in production-constrained spatial-interaction models. *Environment and Planning A* 10: 371-88

Harris, C.C., and Hopkins, F.E. 1972. *Locational Analysis*. Lexington Books, Lexington, Mass.

Miller, O'Dell, and Paul. 1984. *Commercial Market Analysis: Upper James–South Mountain Study*. Hamilton

O'Kelly, M. 1981a. A model of the demand for retail facilities, incorporating multistop, multipurpose trips. *Geographical Analysis* 13: 134-48

– 1981b. Impacts of multipurpose trip-making on spatial interaction and retail facility size. PhD thesis, Department of Geography, McMaster University, Hamilton

Regional Municipality of Hamilton-Wentworth. 1982. *Environmental Assessment Submission*. 2 vols. Hamilton

Scott, A.J., and Storper, M., eds. 1986. *Production, Work, Territory*. Allen and Unwin, London

Social Planning Council of Metropolitan Toronto. 1979. *The Local Property Tax in Metropolitan Toronto.* Toronto

Tabb, W., and Sawers, L., eds. *Marxism and the Metropolis.* 2nd ed. Oxford University Press, New York

Thompson, R.W. 1983. The economy of Metropolitan Hamilton – some basic facts and recent trends. Unpublished paper, Department of Economics, McMaster University, Hamilton

Walker, D. 1978 *The Great Winnipeg Dream.* Mosaic Press, Oakville

Webber, M.J. 1986. Regional production and the production of regions: the case of Steeltown. In Scott and Storper (1986)

CHAPTER 16

McMaster University in a changing urban environment

L.J. KING AND G. OZORNOY

The period of fifty-five years from 1930 to the present, the life of McMaster University in Hamilton, has been one of changing fortunes for both the university and the city. From a small denominational university with an enrolment of fewer than 700 students, a handful of faculty and staff, and four buildings in its early years in the city, McMaster has grown steadily into a major university with more than 11,000 full-time students, more than 6,000 part-time students, around 1,000 each of faculty and staff members, and over 35 main buildings (see figure 16.1). Its teaching and research programs now encompass a wide range of disciplines from the basic arts and sciences to the professional fields of business, engineering, and the health sciences. It ranks today as one of Canada's leading research institutions, and as much as one-third of its total annual revenue is derived from research grants and contracts.

The city of Hamilton too has grown, doubling its own population over the period, and joining now with the surrounding Wentworth County to form a regional municipality that has a total population of well over a half million people. Its economy has expanded apace, although subject from time to time to cyclical changes, and while some diversification has occurred with a growing emphasis on service activities, it has remained nevertheless predominantly a manufacturing city. We shall elaborate on some of these features later in the discussion.

The shared confidence of the leaders of the university and the city in 1930 that the two institutions would benefit one another, has been amply borne out.[1] The city and the region have provided strong and continuing support for the work of the university not only by way of sending their sons and daughters to the university and providing direct financial assistance through tuition fees and municipal grants, but also by way of the dedicated work of its citizenry on university boards and committees. The

Figure 16.1 McMaster University today

university, in turn, makes significant contributions not only to the economic life of the city and region as a purchaser of goods and services, as a major employer, and as a provider of expertise, but also to the social and cultural life through its educational, artistic, and recreational programs.

These relations between a university and the society and community within which it functions are complex and evolving ones and as such they have been the subject of many learned discourses over time.[2] Today, in Canada, they have acquired a special interest in the context of the ongoing debates about the possible paths for the future economic development of Canada. Many commentators from the business world, government, and the academic world are placing particular emphasis on what they perceive to be a critical and essential role for the universities in helping to shape the future economic character of this nation.[3] The discussions today are not as narrowly focused, as they often were in the past, simply on the direct economic benefits that flow from the universities to the communities in which they are located, although these are still of interest, but rather on the broader issues of how the universities might contribute to the changing of the structure of the economy and to the fashioning of the new cultural, social, and political institutions that will be required by such changes. There is considerable emphasis placed in all of these discussions upon the notion of 'technology transfer' from the universities to society as being the key to this heightened importance of the universities, and while the precise meaning to be attached to this notion is not always clear from the discussions, it is seldom difficult to gain a sense of what is envisaged, namely a belief that the universities will serve as the front for those new ideas and technological innovations which, when moved into the world outside of the laboratory and library, will provide the ways to economic salvation. These expectations certainly have been heightened in many people's minds of late by the spectacular successes that have flowed from some of the recent university-industry co-operation in the field of molecular biology, although the cautionary words written about many of these same experiences, unfortunately, are not always heeded.[4]

The situation of McMaster University in Hamilton highlights the central issue addressed in many of these current discussions. The city's economic and social well-being remains primarily dependent on the fortunes of the steel industry and related activities which are today operating in a very difficult economic environment characterized by increasing competition from lower-cost foreign producers, a slow-growth domestic market, and an expanding range of substitutions for their products.[5] If major dislocations in the local labour market with all of their attendant social and political stresses are to be avoided, then some ways must be found to restore the competitiveness of these industries. Technological improvements and innovations represent one such solution and these, in turn, are the products of research such as is done at McMaster. We shall describe in a later section of this essay how this research is organized and how it is transferred to the local industrial economy.

A complementary solution to the difficulties of an urban economy such as Hamilton's is diversification of its economic base, such that there is less dependence on a few activities, especially those which are overly sensitive to economic downswings, and broader dependence on a range of economic activities including ones which have higher growth potential. Of the latter economic activities, it is true to say today that they are knowledge-dependent activities. The corollary of this observation is that we live today in an 'information society' within which education and training must have central roles.

As an introduction to our fuller discussin of McMaster's role in the contemporary setting of Hamilton and the region, we describe briefly what is meant by 'information society' and we comment on Hamilton's situation in the light of that description.

THE INFORMATION SOCIETY

A striking characteristic of western post-industrial society is the importance of knowledge. The social and economic well-being of all such nations today, and Canada is no exception, is dependent on them having a well-educated labour force, access to innovations and discoveries, and the means of rapidly and effectively disseminating information.

Freeman and his colleagues in the United Kingdom have suggested that 'it is by no means unreasonable to link together the electronic computer, semiconductor technology, automation and information technology and to describe them collectively as a new industrial revolution or an "information revolution".'[6] The 'information society' which is emerging as a result, can be thought of as one that is being shaped by three interrelated forces, namely: 1 / the transformation of materials-based manufacturing economies into information-based service economies; 2 / the economy-wide application of information on new technologies in transforming production and distribution processes; and 3 / the development of new ideas, policies, and institutions to manage social issues arising from economic and technical changes.

Most of the available economic evidence on the 'information society' relates to the changing industrial and occupational structure of the labour force. For example, a recent study by the OECD indicated that by 1981 every second North American worker earned his or her livelihood by either producing, processing, or distributing information.[7] During the 1970s in Canada, as Hepworth shows,[8] 80 per cent of the new jobs created were in service industries and 60 per cent of these were concentrated in information occupations. Over the same decade, the proportion of the nation labour force working in services increased from 62.2 per cent to 67.2 per cent, with corresponding declines in the primary and secondary employment sectors. The same patterns of structural change, Hepworth notes, were evident at the regional level such that by 1981 about two-thirds of the labour force in each Canadian province worked in services. Within the service sector, an accelerated growth was observed in those post-industrial services which include information and human services, and their increased prominence is consistent with the model of post-industrial socio-economic development that Daniel Bell, in particular, has formulated.[9]

In western industrial society, the economic health of cities such as

Hamilton derived from their functions as centres of goods-production and -distribution and as gateways in the transportation systems for circulating the factors of production and manufactured goods. In the post-industrial information society, the prosperous cities increasingly function as centres for producing, processing, and distributing information and are interconnected by telecommunications systems. A reflection of some of these changes is seen in the employment pictures of the metropolitan areas of Southern Ontario (table 16.1). Manufacturing employment declined in all four metropolitan centres over the five years from 1980 to 1985, and in the cases of Hamilton and London the net change over the decade was also negative. The gains in manufacturing employment over the decade in Kitchener-Waterloo and Toronto undoubtedly resulted from the greater concentrations in those areas of faster-growth, high-technology industries. With the exception of London, all areas experienced high rates of growth over the decade in the 'services' and the 'finance, insurance, and real-estate' sectors. The 'trade' sector, too, showed increases over the decade in all centres. In interpreting London's performance, it is important to note that it was only one of the four metropolitan areas to show a decrease in total employment from 1975 to 1985 and that notwithstanding the declines in its service sector, that sector still accounted for a higher proportion of its total employment in 1985, 37 per cent, than it did in any of the other three areas.

The recent economic experiences of the North American and West European city have suggested that knowledge and information have become the crucial resources.[10] The most recent recession in this country with its severe levels of unemployment hit hardest those urban economies dominated by the older and heavier forms of manufacturing in which production techniques, labour skills, and management methods have been unchanged over a long period of time. Hamilton is one such city where not only do comparatively high levels of unemployment persist (table 16.2), but where even in the expanding service sector comparatively low levels of average earning prevail (table16.3). It is observations such as these which prompt the gloomy forecasts that in the midst of technological change, certain Canadian cities are home to an expanding low-income, immobile, under-educated, and increasingly isolated population whose problems have become inter-generational.[11] By contrast, those urban areas whose economies are driven by high-technology and post-industrial service activities were less affected by the recent economic downturn; witness the experiences of Toronto and Kitchener-Waterloo (table 16.2).

TABLE 16.1

Percentage changes in employment, by industrial sector, for census metropolitan areas in 1975–80, 1980–5, 1975–85

	Hamilton			Kitchener-Waterloo			London		
	1975–80	1980–5	1975–85	1975–80	1980–5	1975–85	1975–80	1980–5	1975–85
Manufacturing	11.2	−14.1	−4.5	22.9	−7.5	13.7	18.8	−18.3	−2.9
Construction	12.0	6.7	19.5	34.3	−13.8	15.7	−4.6	−15.3	−19.5
Transportation, communications, and utilities	−6.9	38.0	28.4	74.0	−28.7	24.0	22.0	−33.0	−18.3
Trade	14.9	6.0	21.8	27.0	−17.4	5.0	7.2	3.5	11.0
Finance, insurance, and real estate	39.4	−3.0	35.1	11.3	13.9	26.8	7.1	−21.0	15.3
Services	12.2	24.5	39.7	38.0	50.8	108.0	−2.0	−3.0	−5.0
Public administration	−11.5	22.3	8.3	n.a.	−12.9	n.a.	−3.9	−17.6	−20.8
TOTAL	11.4	4.8	16.8	34.7	2.6	38.2	5.8	−10.0	−4.8

SOURCE: Statistics Canada, Labour Force Survey Sub-division, Ottawa. Unpublished data provided upon request.

TABLE 16.2

Unemployment rates in metropolitan areas 1975–85
(% of labour force, annual averages)

Year	Hamilton	Kitchener-Waterloo	London	Toronto	Ontario
1975	6.9	5.9	7.1	5.5	6.3
1976	5.4	5.9	6.7	5.4	6.2
1977	5.9	6.1	5.7	6.6	7.0
1978	6.7	9.0	6.3	6.7	7.2
1979	6.2	7.5	6.4	5.0	6.5
1980	6.4	8.2	7.1	6.4	6.8
1981	6.4	8.1	7.0	4.9	6.6
1982	11.7	10.0	9.4	8.1	9.8
1983	11.9	9.9	9.6	9.0	10.4
1984	9.8	8.5	7.8	7.7	9.1
1985	8.8	7.1	8.8	6.8	8.0

SOURCE: Statistics Canada, *The Labour Force*, Cat. 71-001

TABLE 16.3

Average weekly earnings in manufacturing and services, Hamilton CMA

				% of manufacturing earnings		
	1975	1980	1982	1975	1980	1982
Manufacturing	237.56	378.10	450.31	100	100	100
Services	118.85	155.43	172.56	50.0	47.1	38.3

SOURCE: Statistics Canada, Labour Division, Ottawa. Unpublished data provided upon request.

THE CHALLENGE FOR THE URBAN UNIVERSITY

For McMaster, located in Hamilton, the contemporary economic and social scene offers a many-sided challenge. One side is represented by the demands of the older established industries for new technological innovations that might restore their competitive market positions. Another is represented by the opportunities associated with the development of those post-industrial, knowledge-intensive economic activities which are seen as the spearhead of the transition to a new knowledge-based economy and which are very demanding of research and educated manpower. Another side of the challenge is represented by the deepening plight of the urban disadvantaged, the unemployed and underskilled persons. Their demands, to which the university can respond only in part, are for newer educational opportunities and retraining programs.

The challenge we have described is one that confronts all universities. The growing urgency of the many and often conflicting demands that stem from it, and the difficulties involved in fashioning appropriate responses to these in times of tightened financial resources, are placing considerable strain upon the fabric of the university.

Much of what universities do today, and have done in the past, is universal in nature, namely the pursuit of truth and excellence, and the conservation and transmission of knowledge, both basic and applied. None the less, there have always been tensions present in any university over the relative priorities to be accorded to different activities, to teaching as opposed to research, to the core arts and sciences as opposed to the professional and more applied subjects in the curriculum, and to service activities within the university as opposed to those involving direct participation in the outside community or the worlds of industry, commerce, and government. Ashby made this point in regard to the British universities when he wrote of their 'split personality' as follows:

Round every Senate table sit men for whom the word university stands for something unique and precious in European society: a leisurely and urbane attitude to scholarship, exemption from the obligation to use knowledge for practical ends, a sense of perspective which accompanies the broad horizon and the distant view, an opportunity to give undivided loyalty to the kingdom of the mind. At the same Senate table sit men for whom the university is an institution with urgent and essential obligations to modern society; a place to which society entrusts its most intelligent young people and from which it expects to receive its most highly trained citizens; a place which society regards as the pace-maker for scientific and technological progress. [12]

The same tensions are present in the modern Canadian university and as was hinted at earlier they show signs of deepening as the external demands upon the institution for direct assistance in solving society's current woes increase while at the same time the purse-strings around the government expenditures on which the university is almost totally dependent are being drawn tighter. Whether the tensions will find expression in a drastic restructuring of the university institution itself remains to be seen.

Hurtubise and Rowat, writing in 1970,[13] felt confident in expressing their belief that 'most Canadian universities could be classified as liberal universities' in the sense that Flexner had suggested when he wrote of 'a genuine university' as 'an organism characterized by highness and definiteness of aim, unity of spirit and purpose.'[14] Only the University of Toronto, they felt, was at all close to becoming a 'multiversity' of the type that Clark Kerr had described some years earlier. In his Godkin Lectures of 1963,[15] Kerr had explored the increasingly close relationships that he saw developing between the universities and the great industrial corporations, and his conception of the modern university as a 'multiversity' captured his belief that the two institutions were becoming less and less distinguishable in terms of their objectives and approaches. One obvious corollary was that a principal function of the multiversity became the training of the technicians and managers for those corporations, be they government, industrial, or financial ones.

This conception of the university or multiversity is certainly not a dominant one throughout Canada but there is unquestionably stronger evidence of it today than there was a decade and a half ago when Hurtubise and Rowatt issued their report. Not only has university research become a much larger enterprise involving an annual expenditure by the federal government of $369 million in 1984, but there is growing pressure from the federal government on industry for the latter to contribute an increasing share of these funds and industry is showing it willingness to do so, albeit at a cautious pace. On a different level, universities themselves and faculty members within them have in many cases become entrepreneurs and formed business enterprises; we shall cite some example of this later. Nor has the intrusion of corporate interests been restricted to research and development activities, for instructional programs, too, have been affected. Co-operative degree programs are an obvious example of how the educational experience may be shaped to serve the mutual interests of students, university, and business. None of these observations supports a conclusion that the strengthening relationships between the universities and industry are necessarily bad or erosive of the basic mission of the university; they simply underscore the point that the influence of the business and corporate world on Canadian universities is already quite strong and it is growing.

TECHNOLOGY TRANSFER AND McMASTER UNIVERSITY

Leo Marx has reminded us that the concept of technology itself has today at least three important meanings which he sees as being 'related to each other as if arranged in concentric circles.'[16] At the centre is that meaning which refers to 'the knowledge, skill, and equipment that men use for practical purposes.' Beyond this the term technology is 'expanded to include the forms of bureaucratic organization through which the practical knowledge, the apparatus, and the skills are administered.' And finally, the outer, 'virtually all-encompassing sense of the word ... refers to a hypothetical danger: the virtual domination of the life of an entire society by the mode of thought most conductive to, and therefore seemingly inherent in, the advancement of technology in the narrow sense.'

When we speak of technology transfer and the university we are in effect calling up the second of Marx's meanings and seeking to determine the forms of organization in and around the univesity which will make possible the transmission, administration, and effective utilization of those skills and knowledges developed within its walls. In this determination we must always be careful not to commit the university to courses of action which blind it to the risks inherent in Marx's third meaning of technology, regardless of how hypothetical that danger!

Those organizational forms which exist at McMaster for the transfer of technology, in the broad sense that is mentioned above, are the subjects of the following sections.

Educational Programs

Of the different organizational forms for technology transfer from the university none is more obvious than the teaching activity conducted within and outside the university by its faculty. The students who come to the university and who graduate from its programs are the most valuable contributions that any university can make to society. To the extent that the entering students are drawn from the surrounding city and region and then find employment after graduation in the same areas, then the local impact of the university will be all the more discernible and important.

The tributary area from which McMaster University draws most of its students, more especially those entering undergraduate programs, is one whose boundaries are approximately 80 km (50 mi) distant from the campus. In 1984-5, for example, 91 per cent of all of the full-time regular session students (undergraduate and graduate in all faculties) at McMaster came from within Ontario and, of these, as many as 40 per cent came from the Regional Municipality of Hamilton-Wentworth (table 16.4). The six regional municipalities within 80 km (50 mi) or so of the campus are the homes of 80 per cent of the Ontario students coming to McMaster, and as table 16.4 shows this proportion has been stable over the last four years. The university is seeking to attract more qualified students from outside the province but any future increases in these numbers are unlikely to alter dramatically the major patterns of student flows depicted in table 16.4.

Information on the place of residence of the university's graduates is

TABLE 16.4
Geographic distribution of full-time regular-session students from Ontario
(% of total, 1975–6, 1980–1, 1984–5)

Regional municipality	1975–6	1980–1	1984–5
Durham	1.7	2.1	2.2
York	9.8	11.7	13.6
Halton	11.4	12.1	12.4
Hamilton-Wentworth	43.6	42.0	40.2
Niagara	8.1	7.2	6.1
Peel	4.1	5.2	6.1
TOTAL	78.7	80.3	80.6

SOURCE: McMaster University, *Registrar's Report*, 1975–6, 1980–1, 1984–5

not as easily obtained as that on the initial registrations. The statistics available from the university's Alumni Association suggest that around two-thirds of the 37,500 or so alumni of the university for whom addresses are known reside within 80 km (50 mi) of the campus.

What are the skills that these graduates of McMaster possess and which they are now making use of in their employment and daily lives? A graduate of any one of the many varied degree programs offered within the university can boast not only of a specialized knowledge of one or more subjects such as classics, economics, mathematics, physics, but, more importantly, of a wider sense of what is involved in learning and inquiry. The development and honing of these critical mental faculties is the essence of a university education and the transfer of these skills, along with the specialized knowledge, represents the best that any university can offer to society. The McDonald Commission stressed this same point with its observations that 'very considerable uncertainties confront predictions of specific educational requirements and many, if not most Canadians will have to undertake considerable retraining during their lives. These realities lead Commissioners to emphasize the value of a solid general education – of learning how to learn – so that Canadians may be well equipped to adapt quickly and efficiently to the changing realities of the labour market.'[17]

As was noted above, the educational programs offered by McMaster are many and varied, and even more so today than in the past. In 1974-5, for example, there were some 56 honours and joint honours programs leading to the BA degree and 18 leading to the BSc; by 1984-5 these numbers had increased to 66 and 23, respectively. In the same vein, the number of streams under the Bachelor of Engineering and Management

Program, which was the first of its kind in Canada, increased from 3 to 6 over the same period.

The specialized demands of the technological society have been well served over the years by the training at McMaster of the scientists and applied scientists who are essential for its support and nurturing. In table 16.5 the numbers of graduates in Science, Engineering, and Health Sciences are shown for three different years and it is clear that the numbers of undergraduate degrees (mainly Bachelor of Science, Bachelor of Engineering, Bachelor of Science, Nursing, and MD) have been increasing, as have the numbers of graduate degrees (PhD, Masters) in Engineering and Health Science. The sharp decrease in the number of graduate degrees awarded in Science is a direct function of the declining enrolments in those programs, a trend which it is hoped will soon be reversed.

In addition to the programs in Engineering and Health Sciences, McMaster also offers degree programs in the professional fields of Business and Commerce, Social Work, and Teaching, and in quasi-professional fields such as Physical Education, Computation, Health and Radiation Physics, Labour Studies, and the visual and performing arts. All of these degree programs, whether at the undergraduate or graduate levels, move one or two steps beyond the ambit of the core arts and sciences disciplines in seeking to respond to the different demands for specialized skills and training required in the modern technological society. In the fashioning of these programs, the university is often assisted by representatives of the community at large and there are at McMaster, for example, advisory

TABLE 16.5
Number of graduates in Engineering, Science, and Health Sciences
(by calendar year)

Year	Engineering		Science		Health Sciences	
	Graduate degrees	Under-graduate degrees	Graduate degrees	Under-graduate degrees	Graduate degrees	Under-graduate degrees
1975	70	102	109	263	19	119
1980	67	175	70	265	43	150
1984	79	260	68	378	65	197
% changes						
1975–80	−4.3	+71.6	−35.8	+0.8	+112.6	+26.0
1980–4	+17.9	+48.6	−2.9	+42.6	+51.2	+31.3
1975–84	+12.9	+154.9	−37.6	+43.7	+242.1	+65.5

SOURCE: McMaster University, *Registrar's Report*, 1975–6, 1980–1, 1984–5

groups with such representation in place in Business, Engineering, and Labour Studies. Citizens from the local community also serve on student admissions committees in the Faculty of Health Sciences.

The broad educational role of the university in the local community also finds expression in its offering of part-time degree programs and in the work of its Centre for Continuing Education. At various times in the past, these two functions were combined under one centre at McMaster, the former School of Adult Studies, but today they function more or less separately. The part-time–study degree programs are principally the responsibility of the faculties although it is still considered essential to have a co-ordinator of these activities who works with and assists the faculties in offering the programs. The professional faculties of Business, Engineering, and Health Sciences provide for part-time–study principally at the graduate degree level and then only in the case of Business is there a comparatively large enrolment. The part-time MBA program in that faculty typically has an enrolment of close to 300 students a year, many of whom are registered in a co-operative work-study option that is unique in the English-speaking university-world of Canada. Most of the part-time undergraduate students at McMaster, who numbered over 3,000 in the regular September-to-June session and about 2,600 in the summer school in 1984-5, were registered in degree programs in Social Sciences, Humanities, and Science. It is reasonable to assume that the overwhelming majority of these students, especially those registering in the September-to-June period, were resident within an hour's driving time of the campus. The university does offer some part-time degree courses at off-campus centres in Brantford, Burlington, downtown Hamilton, Hagersville (now discontinued), Oakville, and Stoney Creek but together these accounted for only 631 course registrations in the regular session of 1984-5.

The Centre for Continuing Education at McMaster is responsible for non-degree course offerings leading to certificate or diploma awards. In mounting its programs, which serve mainly local residents, the centre often works in conjunction with external professional organizations such as the Canadian Institute of Management, the Institute of Canadian Bankers, and the Society of Management Accountants. Others of its programs are tailored with particular community groups in mind, which is true of its offerings in labour studies, human services, and metallurgy. Over the past several years the centre has served anywhere between 4,000 and 6,000 students annually and this level of commitment will continue.

The centre is not alone in offering university-based continuing-education programs. The professional faculties of Business, Engineering and

Health Sciences all have some such programs serving the needs of local businessmen, engineers, and health professionals. The McMaster Institute for Energy Studies and the Office on Gerontological Studies also offer continuing-education activities on a more occasional basis.

One special form of educational activity which warrants mention at this point is that of the Small Business Consulting Service run by the Faculty of Business at McMaster. This service, which was formerly identified as the Small Business Advisory Unit, has been operated by MBA students and their faculty advisers for almost fifteen years. The group, which usually numbers around ten or so students, provides technical advice and assistance to small businesses operating mainly in the Hamilton-Wentworth region. It represents a form of university 'out-reach' activity not unlike that of the consulting services provided by individual faculty experts and about which we shall have more to say shortly, but inasmuch as this particular activity involves the work of students who are enriching their own education experience while assisting local businessmen, it is appropriate to include it here under education programs.

In concluding this brief survey of the education activities of McMaster and their relations to the surrounding community, it is as well to recall the point made earlier about the tensions that are ever-present within any university. The evolving demands of the 'information society' are having influence on McMaster University and its curriculum. Some of the impacts are direct and structural, such as the establishment in 1985-6 of a new Department of Computer Science and Systems and an Institute of Molecular Biology and Biotechnology, and there are attendant difficulties and tensions as resource-allocation patterns have to be adjusted to accommodate these new developments. Other impacts are subtler and find expression in the changing curriculum. New programs are proposed such as the ones this year in Civil Engineering and Computer Systems, and Computer Science and Psychology; new demands are made under 'required courses'; and options may be restricted by way of reductions in 'electives.' George Grant, a distinguished scholar and former professor at McMaster, would see in all such changes the 'technological threat' to the university curriculum and a reinforcement of his conclusion that 'the university curriculum, by the very studies it incorporates, guarantees that there should be no serious criticism of itself or of the society it is shaped to serve.'[18] This damning conclusion for the university should not go unheeded, but neither should it go unchallenged. There is much to be improved on in the present university curriculum and as the director of McMaster's innovative Arts and Science Program has observed, 'The university is not the place of learning we want it to be because an over-specialized, compartmentalized, fragmented curriculum does not encourage a search for mastery and understanding.'[19] But there is no going back to the world that Grant longs for, a world freed of the dominance of 'the quantifying and experimental methods.' The challenge rather is to reform the curriculum and, through it, to educate future generations such that they have the requisite skills and aptitudes to live happily in the new information society while feeling free and able to voice their intelligent criticisms of its less-desirable features. It is to that goal that McMaster is committed.

Research Programs

The organization of the scholarly and research activities within a university such as McMaster is the framework within which new knowledge is created and through which this knowledge is transmitted into the world outside of the university.

McMaster University is one of Canada's leading research institutions; in 1984-5, for example, its total sponsored research expenditures of around $50 million were exceeded only by those of universities of Toronto, McGill, British Columbia, and Alberta, all of which are very much larger institutions in terms of the number of students and faculty.[20] Of McMaster's total research budget in that same year, slightly more than 40 per cent was received from the three main federal research councils (the Natural Sciences and Engineering Research Council [NSERC], the Medical Research Council [MRC], and the Social Sciences and Humanities Research Council [SSHRC]).[21]

The heart of the scholarly and research activity within the university is the work of individual faculty members in pursuing topics and questions that are of interest to them. Each faculty member is expected not only to teach and supervise the work of students but to be engaged in scholarship or research and to share, both with students and colleagues, the results of those endeavours. Students, in turn, and especially graduate students, are expected to learn not only by way of attending lectures and seminars but also by preparing essays, projects, and theses. The more advanced the degree program, the higher the level of originality and creativity expected of the apprentice-scholar.

In the humanities especially, the scholarly work typically is carried on by individual scholars and the projects may require many years of painstaking archival work before yielding results in the form of a book. In the experimental and applied sciences, including Engineering and Health Sciences, by contrast, the faculty researcher will work much more closely on a project with a technician or two, some graduate students, and perhaps even post-doctoral fellows. These research projects will typically produce results that can be reported on in several papers a year that are published in the professional journals of the disciplines. In the Social Sciences and Business, the arrangements may be of either form. There are research areas in those faculties where research is conducted in a mode similar to that in the sciences, sometimes with more than one investigator involved and with graduate students contributing to the projects in their own thesis work. In other areas, the approach to research is more akin to the humanities with individual scholars pursuing their work alone.

Whatever the approach and the organization involved, there is in all of this scholarship and research an overriding concern for the search for new knowledge and deeper understandings of different aspects of the physical, biological, social, cultural, or mental world. The contributions flowing from these activities find their way into the wider society outside of the university mainly by way of publications, an important form of technology transfer, which are then evaluated in many ways.[22]

The tendency in certain fields of research today towards team research, and this is particularly obvious in many of the physical and medical sciences, often creates demands for special supporting arrangements in the form of university institutes or centres with their own research facilities and personnel. There are a number of these at McMaster University. The oldest and largest of these organization is the Institute for Materials Research which emphasizes interdisciplinary research on solid materials, from metals, ceramics, and glass to electro-optical devices. It involves the work of more than forty faculty members from as many as nine departments in Science and Engineering, and it employs a dozen or so technical staff who operate and maintain a wide range of specialized experimental facilities. The Communications Research Laboratory is another such institute which brings together the research efforts of ten or so faculty members interested in signal processing, satellites, digital and optical communications, and surface-wave acoustics. The Institute of Polymer Production Technology supports the research of seven full-time faculty members from Chemical Engineering in the area of process technologies for polymer production. The Energy Institute is a smaller centre which aims at fostering interaction between scientists and social scientists engaged in different fields of study concerning energy. Finally, the Institute for Molecular Biology and Biotechnology came into existence in July 1986 with the aim of developing both basic and applied research in molecular biology by scientists drawn from the faculties of Science and Health Sciences.

In these special institutes and centres the interplay between curiosity-driven scholarship and research on the one hand, and the more applied lines of research on the other, which is present throughout the university, comes into sharper focus. For the most part, the institutes grew out of the basic research activities that were being carried on by faculty members. They began as arrangements whereby these research activities could be drawn together to concentrate on research questions that were of mutual interest, and as facilitating organizations which would allow for the sharing of major equipment installations and technical support staff. In the case of the more recently created ones, there was also an explicit acknowledgment of the vital role that such university institutes might play in technology transfer and, not surprisingly, some support for these institutes was sought and obtained from industry. We shall elaborate on this point later in our discussion.

It is perhaps useful at this point to comment briefly on the subject of basic versus applied research, which has been addressed in many discussions of science and science policy.[23] In so far as the distinction relates to what motivates the scholar and researcher and directs their lines of thinking, it remains a crucial one. Basic research may be thought of as being motivated and driven by the desire to know, by a pure intellectual curiosity untainted by any serious concern for whether what is being sought by way of explanation and understanding will be at all useful outside of the university library or laboratory. Applied research, or as it is sometimes called mission-oriented or strategic research, is different in the sense that it is directed towards the solution of some particular problem that is of concern in a setting outside of the university. Typically, most of the scholarship and research in the faculties of Humanities, Science, and Social Sciences at McMaster will be of the basic type; that in the faculties of Business, Engineering, and Health Sciences is more likely to be applied. Such a classification is, of course, an oversimplification, and just as there will be much applied work going on in, say, a department of economics or chemistry, also there will be basic research being conducted in a department of electrical engineering or medicine!

It is also true that the distinction between basic and applied research is today becoming blurred and the time between discovery and commercialization compressed in many existing and emerging fields.[24] Some observers have noted, indeed, that the wedding of science and technology is everywhere apparent.[25]

The contributions of basic scholarship and research cannot, by their very nature, be seen as having a local geographic impact. They are an essential part of any university's contribution to society as a whole, and the local community in which the university is located can claim only the pride of having distinguished scholars as members of its community. Hamilton has shown such a pride towards those many members of the McMaster faculty who have made their mark in the world of scholarship and research.

It is in the domain of applied research that the linkages between the university and the local community are more specific and tangible. They exist in several forms. There is first the consulting activity which involves faculty members in sharing their expertise with local companies and industries in the solution of particular problems. These consulting activities normally are arranged privately between the faculty members and the outside agencies or companies and while the university welcomes such opportunities for its faculty to serve in these ways, it does insist that a code of ethics be honoured.

These consultations often lead to the signing of more formal contractual arrangements between the university and outside clients for the completion of specific research projects. In 1984-5 McMaster University contracted for almost $4 million worth of such research, much of it admittedly with government agencies and companies outside of the Hamilton region, but nevertheless about one-quarter of it with industries located within about 80 km (50 mi) or so of the campus.

University-Industry Co-operation

More direct forms of co-operation between universities and industry have developed in many parts of the world. The Canadian experience in this regard is not nearly as extensive as that of the United States[26] or Great Britain[27] but it is growing.[28] While there are many potential difficulties in the close co-operation of the two sectors, it is generally acknowledged that strong relationships between the two are essential to the 'innovation development cycle,' which is defined as 'a series of stages: technology idea; commercilaizable ideas; productization; marriage of technology with entrepreneur; start of a business; and expansion of a business.'[29]

The relationships that McMaster University has with industry are widening and promising ones. We shall mention three different forms that they take, beyond those which are described above under consulting and contract research.

There is first of all the joint participation of the university and industries in research programs and increasingly in ones that are partially funded through either provincial or federal government arrangements. The Industrial Professorship program which is administered by the Natural Sciences and Engineering Research Council (NSERC) provides an excellent illustration of this form of co-operation. Under the program, McMaster University with assistance from industries located throughout Southern Ontario has been able to date to establish three senior faculty positions, two in communications engineering and one in engineering physics, and others are under discussion. In each case, the senior faculty appointment along with supporting research facilities is funded by a combination of NSERC, industry, and university contributions. The support by industry for senior faculty positions, or chairs as they are oftentimes referred to, is by no means new. At McMaster for example, the Stelco Chair in Metallurgy has been in existence since 1960 and the local industry has provided both salary support and research funds for the holder of the position. Westinghouse of Canada has provided similar support for a chair bearing its name, and more recently a group of local companies helped establish a chair in glass science and technology. What is new and expanding is the scope of both federal and provincial funding arrangements for such university-industry partnerships. NSERC, in addition to supporting the Industrial Professorship program also funds industrial post-graduate scholarship awards and other university-industry exchanges of research personnel. Growing out of its earlier 'Program Research Applicable in Industry' (PRAI) program is the present Cooperative Research and Development Grant program which in 1984-5 provided over $400,000 for such research projects at McMaster. It is especially worth noting that under the former PRAI program and NSERC's other Strategic Research Grants program, a particularly promising line of research into a new steel-making process has been carried on at McMaster in recent years by the holder of the Stelco Chair. This process, for which world-wide patents have been applied for, is about to be tested in a pilot plant. It is perhaps the best recent example of the potential of McMaster research in contributing to the economic development of both the Hamilton region and the nation.

It is clear from the commitments made in 1986 by the federal government to the matching of future funding contributions from industry to university research that the opportunities for strengthening these partnerships will be available.[30] Whether all parties – the universities, industry, and government – will agree on how these opportunities should be capitalized on for the long-term benefit of this nation remains to be seen!

At the provincial level there is also support for university-industry research through the University Research Incentive Fund which is a legacy of the now defunct Board of Industrial Leadership and Development (BILD).[31]

Under the existing program, McMaster has a number of projects being funded which involve the participation of local industries such as Stelco and Westinghouse.

A different form of co-operation between McMaster and industry is seen in the program of corporate memberships in the Institute of Materials Research and the Institute of Polymer Production Technology. The Industrial Affiliate Program of the former allows for six or so companies year to have full access, on the payment of a fee, to the research facilities of the institute. In the case of the Polymer Institute, some sixteen member companies from Canada, the United States, and Europe pay an annual membership fee which provides, at this early stage in the institute's history, the seed money to support applications to both the federal and provincial governments for significant financing of the institute's activities. The member companies are provided with all interim technical reports from the institute and assist in influencing the direction of the institute's research through representation on an industry and government advisory committee.

It was suggested earlier that the final stage in the innovation-development cycle involves the establishment of businesses and their subsequent growth. At McMaster this 'spinning-off' of companies has involved both the university as an active corporate participant and McMaster faculty members as the founders of separate companies. Computer Integrated Manufacturing (CIM), now located in Ancaster, is one company that was spun off by the university. It began as an on-campus, non-profit organization developing computer-aided design and manufacturing processes. Then in 1982 the university incorporated a share-capital commercial company, linked to the continuing 'not-for-profit' company, and with the generous assistance of the Regional Municipality of Hamilton-Wentworth which provided serviced land in the Ancaster industrial park, moved the operation off-campus into a newly constructed facility. The university retained control as the single largest and majority shareholder, with the balance of the common shares being held by Mohawk College and employees of the company. CIM is engaged in technology transfer through the provision of computer-aided design/computer-aided manufacturing (CAD/CAM) expertise to industry on a contractual and fees-for-services basis. Nuclear Activation Services (NAS) is another company in which the university is a partner, only in this case it is a 50/50 partnership between the university and X-Ray Assay Laboratories (XRAL) of Toronto. The company is based on the campus and conducts trace-element analysis, using the McMaster Nuclear Reactor, mainly for the mining industry but also for clients in the health, medicine, and environmental fields.

A different form of spin-off is illustrated by the Statistical Software Group. This company, now located in Hamilton, has an exclusive worldwide licence with Hewlett-Packard Company of the United States to convert and develop statistical and research-oriented data-base management software packages to be used on Hewlett-Packard computers. This work began in the Faculty of Health Sciences, largely on the initiative of a then full-time faculty member, and over the course of seven years developed to a point where its commercialization appeared feasible. The faculty member in question was interested in pursuing this possibility and resigned from a full-time faculty position to head the new company. The university, in return for a 10 per cent ownership of the new corporation and a seat on its board, turned over all existing contracts and customers that had been developed as a result of the related work within Health Sciences.

There are many other instances throughout the university of faculty members incorporating their own companies or assisting in the formation of companies and some of these have been most successful. However, inasmuch as the university does not have an equity position in any of these other ventures, and does not require its faculty members to report the details of such involvements, it would be inappropriate to describe them here. Suffice it to say that the overwhelming majority of them contribute in a postitive manner to the transfer of technology and to the economic development of the region.

TOWN AND GOWN: OTHER IMPORTANT RELATIONSHIPS

The main intention in this essay was to explore some of the meaning of the notion of technology transfer and to suggest how these apply to the relations between McMaster University and the Hamilton-Wentworth Region. We have focused, therefore, on the education, research, and entrepreneurial activities carried on within the university and on the relations these have to the society outside.

There are, of course, other more direct relations between McMaster and the regional municipality that are associated with the roles of the university as an employer, as a major purchaser of goods and services, as a centre for health services and cultural and recreational activities, and as a participant in community affairs. We shall comment briefly on these relations.

Economic Impact of the University

Only certain of the economic benefits and costs associated with a university's presence can be measured directly; others must be inferred. McMaster University, with a total work force of around 5,000 persons, is the third largest employer in the Hamilton-Wentworth Region after the two major steel companies, Stelco and Dofasco. The overwhelming majority of these university employees live in the same region.

In the calendar year 1984 these employees received a total of around $70 million in salaries and wages, excluding benefits. It is reasonable to assume that after taxes and other deductions were made, the total disposable income amounted to about $45 million. If a figure of 10 per cent is used as a rough measure of the propensity to save on the part of these employees, then consumer spending by this group probably amounted to around $40 million during that year. A significant portion of these expenditures would, of course, be made outside of the Hamilton-Wentworth Region but even allowing for this 'leakage,' which is extemely difficult to measure, the direct cash-flow impact upon the local economy was, and continues to be, very significant.

These local expenditures generate, in turn, certain multiplier effects in the form of jobs created. Again, these second-stage effects are not easily determined and it is conventional to estimate them simply by applying a multiplier, usually 2.0, to the total base employment.[32] Hence, the total McMaster-related employment could be said to be around 10,000 jobs.

In addition to direct spending by university employees, there is a significant amount of spending by the students. Recall that McMaster currently has enrolled over 11,000 full-time students, of whom 2,400 live on campus in the student residences and an estimated 4,500 live at home. The advice given these days to potential students from outside of Canada is that each person normally will require at least $6,000 to $7,000 for expenses beyond those of university tuition fees. This observation, together with those on the proportion of McMaster students living at home and in the university residences, suggests that a reasonable estimate of the average annual expenditure per student might be around $4,000 and of the total annual expenditure by McMaster students, $44 million. Again, some portion of this spending will be channelled outside of the local region. These estimates of student expenditures should for completeness' sake be supplemented with the corresponding ones for conference visitors, summer-school students, and so on. Unfortunately, such numbers are not readily obtained and all that can be said is that there are sizable economic gains for the local community resulting from such activities.

There is a further important economic impact of the university on the local region resulting from the university's purchase of goods and services. McMaster annually spends now over $7 million on 'supplies and expenses' and almost $1 million on 'renovations and maintenance.' It is impossible to say precisely how much of this money is spent locally but the suggestion made in an earlier study of this issue to the effect that it is at least half still seems reasonable.[33]

A final comment on the economic impact of the university relates to the overall balance sheet that is involved. In the early 1970s several studies in the United States sought to address the question of whether a city's loss in property tax revenue from the presence of a university within its boundaries was offset by the benefits gained.[34] The conclusion drawn from these analyses were always the same: the economic benefits generated by the university more than compensated for the hypothetical amount that the city would receive in taxes if some other form of enterprise were located on the property occupied by the university. Given the widespread recognition that exists these days, especially in Ontario, of the vitally important economic roles that the universities play in the communities in which they are located, it is hardly surprising that such studies seem to be no longer in vogue.

Health, Cultural, and Recreational Activities

In these domains, the influence of McMaster on the surrounding community is not directly measureable but it is real. The presence of the large Faculty of Health Sciences, for example, has had strong effects on the organization, the character, and undoubtedly the quality of health-care delivery arrangements in the region but it would be impossible to gauge with any accuracy the overall impact of these effects on the levels of community health. There are simply too many other complex and highly interrelated social, economic, environmental, and nutritional factors at work for this to be accomplished. But it can be said with utmost confidence that the Faculty of Health Sciences and its members are an extremely valuable resource within the community. Its five hundred or so full-time faculty members are not only responsible for educational and research programs similar to those offered in the rest of the university, but more than half of them are also engaged actively in clinical work involving patients from the community. These faculty-physicians may be based on campus, at the Chedoke-McMaster Hospital, or in one of the other Hamilton hospitals. These various hospitals along with McMaster and Mohawk College are represented on the Hamilton-Wentworth District Health

Council which is responsible for planning and developing the health-care services and systems of the region. Faculty members from Health Sciences are also active participants in various local committees and organizations concerned with health-related issues such as occupational health and safety, environmental pollution, and toxicology.

The quality of community life in the region is enhanced also by the cultural activities of McMaster University. The concert program offered by the Department of Music which many local citizens regularly attend and the public performances by the McMaster Symphony Orchestra and the University Choir contribute to such enrichment. The university's Art Gallery and its libraries are also accessible to interested citizens. The public lectures by distinguished scholars and the periodic university open-houses afford further opportunities for closer interaction between town and gown.

Local athletic clubs and schools are welcome of many of the university's athletic facilities and the Hamilton-Wentworth Region has contributed generously to the recent rebuilding of the outdoor track. The Fitness Camp program run by the university in the summer months attracts 450 local school-children as participants and provides employment for forty senior university students.

Community Service

As citizens of the Hamilton-Wentworth community, university faculty and staff members have served from time to time on bodies such as school boards, planning committees, and other public service organizations. They have run for public office and been successful at it: witness the fact that two of the provincial ridings within the City of Hamilton are at present represented at Queen's Park by persons on leave from McMaster.

The participation has not been a one-way street and members of the community regularly serve on committees and boards within the university. Over the years, McMaster has been particularly well served by a succession of industrial leaders, local businessmen, and professional people who have held seats on its board of governors, its senate, and the major committees of these governing bodies. Their dedicated service and wise counsel has helped build the strength of the university. In the professional faculties of Business, Engineering, and Health Sciences many of the same interested citizen and others have assisted through service on advisory councils and committees.

These patterns of interaction are an expression of the common interest that binds the university and the region together. This sense of a shared destiny was the driving force behind the successful Economics Futures Conference which the university and the Hamilton Spectator Newspaper Company co-sponsored in 1983 in Hamilton.

CONCLUSION

The partnership between McMaster University and Hamilton-Wentworth that has been forged over the past fifty years is now a strong and flourishing one and there is every reason to suppose that further benefits will result from it. The economic and other impacts of the university certainly will be ever present and the cash-flows will remain significant.

We have tried to suggest, however, by the emphases in this essay that there are even more important impacts by way of technology transfer, in all of its dimensions, which will only be rewarding and continuous if it is driven by first-class teaching and research activities within the classrooms, libraries, and laboratories of the university. The challenge to ensure that this working environment within the university remains attractive to the future generations of bright young scholars, scientists, and engineers is one that we believe must be addressed seriously by all groups within Canadian society.

NOTES

1 See Charles M. Johnson, *McMaster University*, vol. 1, *The Toronto Years*, Chapter 10, and vol. 2, *The Early Years in Hamilton, 1930-1957*, Prologue, Chapter 1 (Toronto: University of Toronto Press 1976, 1981).

2 See, for example, Eric Ashby, *Adapting Universities to a Technological Society* (San Francisco: Jossey-Bass Publishers 1974); Laurence Stone, ed., *The University in Society*, vols. I, II (Princeton: Princeton University Press 1975); Nicholas Phillipson, ed., *Universities, Society, and the Future* (Edinburgh: Edinburgh University Press 1983); Burton R. Clark, ed., *Perspectives on Higher Education* (Berkeley: University of California Press 1984).

3 For example, D. Vice, 'Post-Secondary Education in Canada: A Capital Investment,' Submission to Senate Committee on Finance, Ottawa, 1986; Royal Commission on the Economic Union and Development Prospects for Canada, *Report*, vol. 2, Chapter 18 (Ottawa: Ministry of Supply and Services 1985); The Commission on the Future Development of the Universities of Ontario, *Ontario Universities: Options and Futures* (Toronto: Queen's Printer 1984).

4 See, for example, N. Wade, 'Background Paper,' in Twentieth Century Fund Task Force, *The Science Business* (New York: Priority Press 1984), 19-84; also Charlotte Gray, 'Taking Research beyond Laboratories: Exploiting Commercial Potential,' *Canadian Medical Association Journal* 134 (1986): 817-24.

5 See, J.R. D'Cruz and J.D. Fleck, *Canada Can Compete! Strategic Management of the Canadian Industrial Portfolio* (Montreal: Institute for Research on Public Policy 1984), Chapter 8.

6 C. Freeman, et al., *Unemployment and Technical Innovation* (Westport, Conn: Greenwood Press 1982), 119

7 OECD, 'Information Activities, Electronics and Telecommunications Technologies,' OECD ICCP Series, no. 6 (Paris: OECD 1981). See also M. Porat, *The Information Economy: Definition and Measurement*, Special Publication, 77-12 (1) (Washington: U.S. Dept. of Commerce, Office of Telecommunications 1977).

8 M. Hepworth, 'Spatio-economic Impacts of the Use of Information Technology by Multi-Locational Organizations,' *Urban Studies*, forthcoming

9 Daniel Bell, *The Coming of Post-industrial Society* (New York: Basic Books 1973); and 'The Social Framework of the Information Society,' in M.Dertouzos and J. Moses, eds., *The Computer Age: A Twenty Year View* (Cambridge, Mass: MIT Press 1978)

10 See B. Bluestone and B. Harrison, *The Deindustrialization of America* (New York: Basic Books 1982); H.J. Ewers, H. Matzerath, and J.B.Goddard, eds., *The Future of Metropolis: Economic Aspects* (Berlin: de Gruyter 1986).

11 National Council of Welfare, *Poverty Profile 1985* (Ottawa: Ministry of Supply and Services 1985)

12 E. Ashby, *Technology and the Academics* (London: Macmillan 1963), 69-70

13 R. Hurtubise and D.C. Rowat, *The University, Society and Government* (Ottawa: University of Ottawa Press 1970), 22

14 A. Flexner, *Universities: American, English, German* (New York: Oxford University Press 1968), 178

15 Clark Kerr, *The Uses of the University* (Cambridge, Mass: Harvard University Press 1963)

16 Leo Marx, 'Technology and the Study of Man,' in W.R. Niblett, ed., *The Sciences, the Humanities and the Technological Threat* (London: University of London Press Ltd. 1975), 7-8

17 Royal Commission on the Economic Union and Development Prospects for Canada, *Report*, 2: 746

18 G. Grant, 'The University Curriculum and the Technological Threat,' in Niblett, *The Sciences, the Humanities and the Technological Threat*, 34

19 H.M. Jenkins, 'On the "Two Cultures" and the University as a Place of Learning,' in McMaster University, *Science and Humanism: Is the House of Enquiry Divided?* Humanities Lecture Series, 1981, 117

20 Statistics are reported in Statistics Canada, *Financial Statistics of Universities and Colleges 1984-85* (Ottawa: Canadian Association of University Business Officers, 1985), Report 2.2c.

21 The Commission on the Future Development of the Universities of Ontario (the 'Bovey Commission') drew attention in its 1984 report, *Ontario Universities: Options and Futures*, to the fact that McMaster derives about 19 per cent of its total operating expenditures from awards made by these three councils; the next higher percentage in the province was for the University of Toronto at around 16 per cent.

22 The publication of book reviews is one established practice. Citation indexes provide a different form of evaluation; in these authors are listed along with the number of

citations to their published papers or books that have appeared in other publications. A more novel approach, that of 'bibliometrics,' was used in 'A Comparison of Scientific Research Excellence at Selected Universities in Ontario, Quebec and the United States, 1982,' an unpublished report prepared by the IDEA Corporation for the Bovey Commission, Toronto. In that report, McMaster ranked seventh out of twelve universities in terms of the 'average influence of scientific research papers' published but was second to Toronto among the Canadian universities (which included, in addition, McGill, Queen's, Western, Waterloo, and Guelph). In 'Medical' papers, however, McMaster led the Canadian group but was sixth overall; in 'Biology' McMaster was third overall behind only Queen's and MIT; and in Engineering-Technology it was fourth overall behind Chicago, Cornell, and Toronto.

23 See, for example, L.P. Bonneau and J.A. Corry, *Quest for the Optimum. Research Policy in the Universities of Canada* (Ottawa: Association of Universities and Colleges of Canada 1972), Chapter 3.

24 B.J. Culliton, 'The Academic-Industrial Complex,' *Science* 216 (May 1982): 960-2

25 F. Seitz, et al., 'Prospects for New Technologies,' in *Outlook for Science and Technology: The Next Five Years* (San Francisco: W.H. Freeman and Co. 1982)

26 See the examples described in A.L. Frye, ed., *From Source to Use. Bring University Technology to the Marketplace* (New York: American Management Association 1985). Also see L.G. Johnson, *The High-Technology Connection: Academic / Industrial Cooperation for Economic Growth*, ASHE-ERIC Higher Education Report no. 6 (Washington, DC: Association for the Study of Higher Education 1984); M. Bullock, *Academic Enterprise, Industrial Innovation and the Development of High Technology Financing in the United States* (London: Brand Brothers and Co. 1983).

27 See, for example, Segal Quince and Partners, *The Cambridge Phenomenon* (Cambridge: Segal Quince & Partners 1985); S.L. Bragg, 'Development of University / Industry Partnerships in Britain,' in *Technological Innovation: University Roles* (London: The Association of Commonwealth Universities 1984), 187-95.

28 See J. Maxwell and S. Currie, *Partnership for Growth. Corporate-University Cooperation in Canada* (Montreal: The Corporate–Higher Education Forum 1984); J.V. Raymond Cyr, *Spending Smarter. Corporate-University Cooperation in Research and Development* (Montreal: The Corporate–Higher Edudation Forum 1985).

29 Johnson, *High-Technology Connection*, 8

30 In announcing, on 26 February 1986, its funding for the three federal research councils, the federal government pledged to match any investment by the private sector up to a maximum of $369 million over the next five years. The commitment involves a 'one for one' matching by government up to a maximum of 6 per cent of each council's annual budget.

31 In the 1986 Speech from the Throne a new commitment was made to 'support, complement and encourage science and technology research in the private sector and post-secondary institutions.' For this purpose, a $1 billion special technology fund was announced, of which at least half was to be new monies.

32 Harvey Schwartz, *Guide to Estimating Employment Multipliers* (Ottawa: Ministry of Supplies and Services 1982)

33 J.A. Cleworth, E.C. Higbee, and S.F. Semeniuk, *The Economic Impact of McMaster*

University on the City of Hamilton and Surrounding Localities (Hamilton: McMaster University Office of Institutional Research 1973)

34 W.A. Strang, *The University of the Local Economy* (Madison, Wisc: Bureau of Business Research and Service, University of Wisconsin 1971); J. Caffrey and H.H. Issacs, *Estimating the Impact of a College or University on the Economy* (Washington, DC: American Council on Education 1971)

CHAPTER 17

Hamilton today

W.G. PEACE AND
A.F. BURGHARDT

Throughout its history, Hamilton has had to contend with the dominance of Toronto, Canada's foremost metropolitan centre. Weaver (1982) notes that: 'No Canadian city has had to endure comparable rivalry from a nearby metropolis' (p. 194). Despite this enormously inhibiting factor, recent changes in the social, economic, and cultural character of the city have served to make Hamilton's future prospects very attractive. The purpose of this essay is to describe these changes, some of which are unique to Hamilton, others of which represent commonalities shared with a broad spectrum of North American cities. The first part of the essay examines the overall growth of the city in terms of population change and physical extent from 1951 to the present prior to focusing on specific developments in the inner city. Following this, various political, cultural, and economic developments of the Hamilton region are highlighted. It is apparent that the cumulative effect of these changes has been to place Hamilton on the verge of a new era.

PHYSICAL EXPANSION AND POPULATION CHANGE (W.G.P.)

In examining patterns and processes of the past three and one-half decades it is revealing to consider first the dynamics of urban change with reference to the physical expansion and population growth of the city. Many of the changes described in this essay are, in part, a response to these overall city and region-wide trends. As illustrated in table 17.1 the Census Metropolitan Area (CMA) of Hamilton nearly doubled its population between 1951 and 1981. However, the rate of increase declined in each of the five-year intervals in question. The city proper has also experienced an absolute increase in total population of 42.7 per cent over the thirty-year period with the rate of increase becoming progressively smaller until 1976.

TABLE 17.1
Population change in Hamilton and region

| | Hamilton | | | CMA | |
Year	Population	% change		Population	% change
1951	208,321	–		281,901	–
1956	239,625	15.0		341,513	21.1
1961	273,991	14.3		401,071	17.4
1966	298,121	8.8		457,410	14.0
1971	309,173	3.7		503,122	10.0
1976	312,003	0.9		529,371	5.2
1981	306,434	−1.8		542,095	2.4

SOURCE: Statistics Canada

Between 1976 and 1981 the city experienced an absolute decrease in numbers. This decrease was the city's first since a 1.4 per cent decline between 1931 and 1936. The slowing of growth in both the city and the CMA has been the result of three general factors. First, surrounding regions and urban centres have been more aggressive (and, therefore, more successful) in attracting new growth. Second, general economic and demographic trends at the provincial and national level have resulted in an overall decline in economic growth. Finally, the relative decline of the manufacturing sector of the economy has hit Hamilton especially hard, given its bias towards heavy industry.

As a response to the population increases in both the city and the CMA, Hamilton has expanded physically from 1951 to 1981. Almost all of this physical expansion occurred between 1952 and 1961 (table 17.2). During this nine-year period the physical extent of the city nearly doubled, largely as a consequence of the city's annexation of Barton and parts of Saltfleet and Ancaster. Subsequent physical expansion has been marginal, with

TABLE 17.2
Physical expansion of Hamilton

Year	Total acreage
1952	16,831
1960	29,519
1961	31,566
1962	31,724
1965	31,597
1966	34,842
1971	30,348
1981	33,464

SOURCE: Weaver (1981): 199

most of the increased acreage resulting from the infilling of Hamilton harbour for purposes of industrial expansion. The city has, for all intents and purposes, reached its limit of physical growth unless it annexes land from the smaller surrounding municipalities. This, however, is highly unlikely.

The combined changes in the population and physical extent of the city described above give rise to certain changes in the tenure characteristics of the city's population (table 17.3). The number of inhabited dwellings increased by 105 per cent from 1951 to 1981. During this same period the average number of occupants per dwelling decreased by 1.1 persons while the proportion of the city's dwellings which were owner occupied declined by 8.7 per cent. These changes in the physical and social environments of the city reflect the recent demographic trend towards smaller families as well as the apartment construction boom of the 1960s. It should also be noted that the composition of Hamilton's labour force has changed significantly over the past three decades (table 17.4). In particular, manufacturing activity accounted for more than one-half of the labour force in 1951 but by 1981 this sector accounted for slightly less than one-third of all employment. Over the same period the industry group showing the largest increase in employment share was the community, business, and personal-services sector.

With this overall picture of Hamilton's growth in mind, it is both interesting and instructive to examine the myriad of changes which have taken place in the central area of the city. In spite of the fact that much of the population increase has occurred in the peripheral areas, it is the inner city which has undergone the most profound metamorphosis. The terms 'inner city' and 'central city' will be used interchangeably to refer to the area bounded by the bay on the north, the escarpment to the south,

TABLE 17.3
Changes in Hamilton's occupancy pattern, 1951–81

Year	No. of inhabited dwellings	Avg. no. of occupants/ dwelling	% owner occupied
1951	55,340	3.8	65.0
1961	73,829	3.7	69.3
1971	94,590	3.3	57.9
1981	113,916	2.7	56.3

SOURCE: Weaver (1981)

TABLE 17.4
Labour force by industry, Metropolitan Hamilton, 1951–81

Industry group	1951		1961		1971		1981	
	No.	%	No.	%	No.	%	No.	%
Primary								
industries	1,698	1.5	4,501	2.9	3,935	1.8	5,380	1.9
Manufacturing	59,553	52.2	61,090	40.3	73,295	34.5	89,320	32.0
Construction	7,246	6.4	10,585	7.0	13,390	6.3	16,285	5.8
Transportation,								
communication,								
and utilities	6,939	6.1	9,374	6.2	10,380	4.9	14,770	5.2
Wholesale,								
retail trade	16,240	14.2	24,028	15.8	32,195	15.1	46,290	16.6
Finance, insurance,								
and real estate	2,775	2.4	4,969	3.3	8,410	4.0	13,630	4.9
Community,								
business, and								
personal services	15,242	13.4	28,830	19.0	49,520	23.3	78,770	28.3
Public								
administration								
and defence	3,385	3.0	5,438	3.6	7,960	3.7	11,045	3.9
Unspecified	905	0.8	2,822	1.9	13,575	6.4	3,255	1.4
TOTAL	113,983	100.0	151,637	100.0	212,660	100.0	278,745	100.0

SOURCE: Statistics Canada

Queen Street to the west, and Wellington Street to the east. This, coincidentally, was the approximate extent of the city in 1876, as discussed in chapter 8.

HAMILTON'S EVOLVING INNER CITY

Much of the written history of Hamilton has focused on the city's economic development, emphasizing the roles of industry and labour in the emergence of Hamilton as a major Canadian urban centre. This economic development has, in part, been responsible for many significant changes within sub-areas of the city. Of particular interest are those physical changes which have taken place in the inner city. It is in this part of the city's built environment that one finds monuments to the city's past next to the cumulative impact of the modern forces of urban development. Three primary forces underlie the changes described here. These are: 1 / the process of suburbanization which contributed to inner city decline;

2 / the impetus for revitalization provided by bold urban renewal schemes; and 3 / the subsequent involvement of the private sector in revitalizing institutions, businesses, and residences alike.

Physical conditions in Hamilton's inner city have varied considerably through the years. The *Canadian Illustrated News* (5 August 1871) offered the following description of Hamilton: 'The "ambitious city" can boast several very fine streets, but King Street, running east and west, and James Street which intersects it at right angles, are by long odds the finest whether as to spaciousness, business importance or architectural adornment.' McKay (1967) notes that the city's economic development in the latter part of the nineteenth century was reflected in the many examples of magnificent commercial, institutional, and residential architecture. The built environment was, therefore, a reflection of the city's prosperity at the turn of the century.

Over the next seventy-five years, however, the inner city experienced substantial deterioration despite the economic growth of the city as a whole. Picturesque descriptions such as that noted above were no longer appropriate. A 1958 survey of Hamilton identified nine areas as being in need of extensive redevelopment. A newsletter issued by City Hall in 1969 (just prior to urban renewal) noted that Hamilton's once proud business district was in a state of continuous decline with substandard buildings calling for long overdue action. More than 80 per cent of the buildings in the downtown area were of pre-1900 vintage (Freeman, 1976). Nader (1976) notes that 'the downtown area [had] a rather rundown appearance, typified by the functional and physical obsolescence of many of the commercial buildings. There [had] been relatively little private redevelopment, mainly because of a lack of demand for commercial office space, which in turn is a reflection of the city's subordinate role in relation to Toronto' (p. 261). Thus, Hamilton's inner city had experienced a significant decline over the first seventy years of the twentieth century.

An aggressive and ultimately highly successful attempt to reverse the process of decline began in the late 1960s with the city's participation in the urban renewal programs under the auspices of the National Housing Act. In particular, three renewal projects were responsible for the beginning of a new era for downtown Hamilton. These projects involved the construction of low-income housing in a rundown residential area on James Street immediately north of the central business district (CBD); the removal of blighted residential and commercial land uses along York Street; and, most importantly, the beginning of a new civic square. Construction of Phase I of the third project (known as Lloyd D. Jackson Square) consisting of office, commercial, and retail space began in late 1970 and was completed in 1972. The Stelco Tower, which was part of Phase I, became (temporarily) the tallest structure in the city. This project represented the first step towards an extensive transformation of Hamilton's downtown.

The list of improvements and changes which have taken place since the completion of Phase I of the civic-square project is lengthy. Over the past fifteen years more than $300 million dollars have been invested in the complex covering more than 16 ha (40 ac) of prime downtown property. The various components of this urban renewal scheme (and their completion dates) include the theatre/auditorium complex Hamilton Place (1973), the Robert Thomson Building (1977), the Hamilton Art Gallery (1977), the new Hamilton Public Library (1980), the new Market (1980), the Hamilton Convention Centre (1981), the Ellen Fairclough Building housing the provincial government offices (1982), the Standard Life Centre (1984), the new Sheraton Hotel (1985), and most recently the arena-trade centre known as the Copps Coliseum. In addition to these projects involving public investment a number of other prominent new commercial and office buildings have contributed to the evolution of the downtown. Included among these are the Terminal Towers shopping and office complex (1966), the Century 21 Building (1973), and First Place (1974). Construction is currently underway on the new, twin-towered Canadian Imperial Bank of Commerce Building at the key intersection of King and James streets. A significant number of smaller scale developments have also added to the improved image of Hamilton's downtown landscape.

Accompanying this boom in office, commercial, and institutional development has been the increasing diversity of related services such as hotels and restaurants. The downtown now boasts three major hotels: the recently completed Sheraton Hotel; the Holiday Inn associated with Terminal Towers; and the recently refurbished Royal Connaught, one of the city's oldest and finest hotels. (It should be noted that the presence of these large chain hotels has contributed to the demise of a number of smaller, private hotels, among them the Fisher Hotel and the Wentworth Arms Hotel, the former being wiped out by urban renewal and the latter being destroyed by a spectacular fire.) Downtown Hamilton also offers a much wider range of popular dining establishments than it did previously.

An equally important facet of change in downtown Hamilton has been the preservation and refurbishing of architecturally and historically significant buildings for commercial and business purposes. Among the most noteworthy examples are the refurbishing of the exterior of the Treble Hall on John Street North, the opening of Park Place (formerly the Right

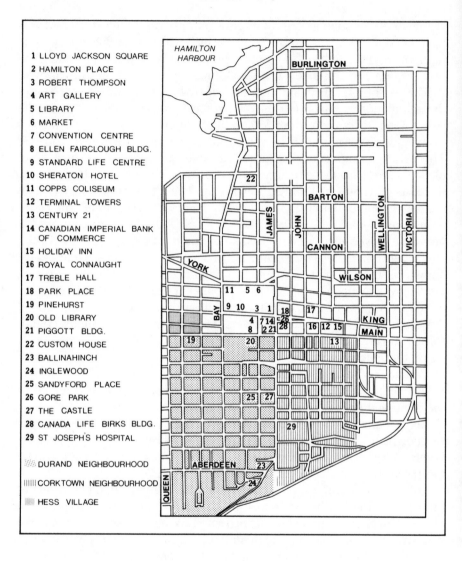

Figure 17.1 A walking tour of downtown Hamilton

House) on King Street East, and the renovation of historic Pinehurst on Jackson Street by CHCH-TV. In addition there have been recent announcements regarding the conversion of the old Library on Main Street West into the new Unified Family Court, the continuation' of renovations to Hamilton's first skyscraper (the Pigott Building on James Street South), and the proposed transformation of the 1858 Custom House on Stuart Street into a performance centre for Hamilton's professional theatre company. (See figure 17.1 for locations.)

Another important aspect of private rehabilitation involves the preservation of architecturally significant buildings for residential purposes. Of particular interest are examples of historic buildings which have been converted into luxury condominiums. One of these (Ballinahinch), located at the base of the escarpment on James Street South, received an award from the Canadian Housing Design Council in 1983 for having 'successfully inserted condominium apartments within the envelope of one of Hamilton's most historic houses without destroying the fine architectural elements and details of either the interior or exterior' (Hamilton *Spectator*, 12 November 1983). Inglewood, an impressive Gothic Revival structure situated on the escarpment plateau on James Street South, and Sandyford Place, an excellent example of Georgian terrace housing, are further examples of the conversion of historic residences into luxury condominiums. In fact most central-area neighbourhoods are experiencing varying degrees of private residential upgrading. This trend is particularly strong in the Durand, Corktown, and Kirkendall neighbourhoods. Several areas within the inner city have experienced upgrading for commercial and office purposes as well as for residential land use. Most notable among these areas are Hess Village, James Street South, and the area adjacent to St Joseph's Hospital. Much of the recent successful preservation and rehabilitation of the city's historic building stock is due to changing attitudes towards the past, brought on in part through agencies such as the Durand Neighbourhood Association and the Local Architectural Conservation Advisory Committee.

While the picture painted in the preceding paragraphs is a positive one, it is worth noting that there have been various examples of negative consequences arising from recent inner-city changes. Some historic buildings have been demolished. Among these are the former Canada Life Assurance Company/Birks Building and (more recently) the Canadian Imperial Bank of Commerce Building. Other important structures have undergone a more gradual deterioration. F.J. Rastrick's masterpiece 'The Castle' at the corner of James Street and Duke Street 'stood originally amidst large grounds, well above street level, and contained in a stone wall' (McKay, 1967: 18). Over the years 'The Castle' has been subjected to a series of degrading alterations. Its current condition is 'a graphic instance of willful ruination of the past' (McKay, 1967: 18). In addition, many fine Victorian houses gave way to the apartment construction boom of the late 1960s and early 1970s. The commercial redevelopment of the central city has also resulted in the loss of many smaller businesses. The recent shift of the CBD to the west of James Street has also brought on a decline in the economic viability of commercial activity on King Street East.

The net result of the changes described above has been the radical transformation of Hamilton's central city for the better. This part of the city (and consequently the city as a whole) is currently much more economically, socially, and culturally viable than it was a mere fifteen years ago. Downtown Hamilton stands as evidence that it is possible to reverse the forces of urban decline through the combination of public and private investment in the built environment to produce an exciting collage of the old and the new. As a distinct geographic area, Hamilton's inner city features a unique blend of the historic and modern forces which give character to the city and surrounding region.

THE CHANGING LIFE OF THE CITY (A.F.B.)

The reconstruction of the city core and the preservation and renovation of the city's older structures are not only significant changes in the city fabric, but also local indications of a revolution in society's evaluation of what constitutes a satisfactory urban experience. Our perception of the role and function of a city has been turned around during the past thirty years, and probably no other city of Canada has gained as much from this change as has Hamilton. Three decades ago, in the time of Louis St Laurent and Dwight Eisenhower, nature was still perceived as a commodity within the urban setting, belching smokestacks were viewed as symbols of wealth and power, and the boring expanses of suburbia were accepted as the ultimate in living style and grace.

Since the late 1960s a new creed for the meaning of urban living has been enunciated and, on the whole, accepted. Five key elements have constituted this new vision: 1 / a renewed appreciation of the social and cultural significance of the city centre; 2 / the belief that the achievement of individual and community self-identity requires the preservation of the better structures of our heritage; 3 / the high value placed on human diversity; 4 / the perception of the older industries as ugly and of the pollution they have spawned as execrable; and 5 / the realization of the great importance of providing easy access to natural greenery and to outdoor sports. Hamilton, which had boasted of being Canada's industrial centre par excellence, had largely ignored its unsurpassed girdle of natural features. The city thus stood to gain enormously from the new vision, and gain it has.

The redevelopment of the downtown has occupied the attention of the city planners and administrators for over twenty years. It must be judged a success in that it has given the city a visible heart and has kept people coming downtown, providing a contrast to the disastrous situation in the downtowns of many cities in the United States. The new facilities have placed Hamilton on a par with any city of its size in Canada, and above most cities its size in the United States. It can be said too that the whole is greater than the parts, in that the collection of commercial and administrative buildings, joined with the high-rise apartment clusters to the southwest, has given Hamilton the look and feel of a modern city. If it is to be taken seriously, a city must look like a city; it must be able to attract to its downtown those affluent couples and singles who have become the prime support of restaurants and many cultural events. The clustering in downtown Hamilton is impressive; for anyone living near the city hall, the theatres, concert halls, art gallery, coliseum, shopping, and restaurants are all within a kilometre's walk.

Of course all of this has had its price. The downtown mall, Jackson Square, has been a great financial success. But the delays in its construction and the policy of its rentals led to the massive replacement of shops run by local merchants with branches of international chains. More troubling yet has been the decay of the eastern portion of the downtown, occasioned by the steady expansion to the west of Jackson Square. The two halves of the King Street shopping belt have not been integrated, and in the winter, sidewalk shopping cannot compete with warm, indoor shopping. Fortunately, a few specialty 'generators,' notably German, have continued to anchor the eastern end of the old CBD. Many attempts have been made to revitalize the eastern end but to little avail so far. The enormous pull of Jackson Square, plus the move of the offices and plant of the Hamilton *Spectator* out of the downtown, and a disastrous bus strike a few years ago have made the continued viability of the eastern shops doubtful.

The construction of the highrise apartment blocks just to the south of City Hall was necessarily at the cost of the loss of older structures. Many fine nineteenth-century homes were levelled. Luckily, the superb mansions at the foot of the escarpment were not touched, and the increased land values in the area led to the renovation and beautification of a number of fine old town houses. The perceived threat of 'developers' obliterating a neighbourhood led to the formation of a vocal, hard-working preservationist organization. Both their efforts, and the appreciation of heritage features which they roused, became important factors in the formation of the local vision of an ideal city.

The continued vitality of the city centre, coupled with the physical constraints of the site, has led to an increase in traffic rushing through the east-west corridor between the escarpment and the bay. Inevitably, but also unfortunately, the traffic engineers have resorted to the expedience of designating certain streets as one-way thoroughfares. Four of these – Cannon, Wilson, King, and Main – run parallel within a span of half a kilometre. The truck traffic which roars along these arteries, as well as several cross routes, has had a severely negative impact on the less-affluent living areas to the east of the city centre. Until now the arrival of new waves of immigrants has maintained the older urban fabric to the north and east of the downtown but, unless some measures are taken to remove the worst of the traffic from these streets, it is doubtful that this east-west expanse near the heart of the city can remain as an acceptable residential area.

It is somewhat paradoxical that the achievement of the modern city should include both the construction of the new and the preservation of the old. To a large extent, the latter was a response to the former. But, beyond traditionalism, the preservationist activity was an aesthetic response; we are as we are because of what our forebears were, and the beauty they fashioned should not be destroyed needlessly. Further, a city should be able to show its children concrete remembrances of the city it was, during the various stages of its growth.

A strong preservationist movement has now been active in Hamilton and several of its suburbs for over twenty years. Overall, one would have to conclude that its efforts have been successful, although it may well be true that the blunting of the runaway highrise construction process was as much a result of the imposition of rent controls as of the efforts of the preservationists. Hamilton is still fortunate to possess, along its James Street axis and flanking streets, a wealth of fine old churches, a historic cathedral, classic examples of an armouries and an Orange Lodge facade, several nineteenth-century office blocks, and dozens of mansions from the

'Gilded Age.' The James Street North area is somewhat run down, but plans for its renewal are underway.

A unique form of preservationism has been the prolonged and difficult project to salvage two American warships which sank in Lake Ontario during the War of 1812. Somewhat fortuitously one of the vessels was named the *Hamilton*, giving the city an added impetus for making the salvage attempt. The '*Hamilton* and *Scourge* Project' promises to be extremely challenging and expensive. The ships will have to be raised from the lake bed without having them fall apart, and then housed under conditions which will prevent deterioration. A special museum to house the two vessels is proposed for the lakeshore. If this project can be completed, the city and Canada will have gained a tourist asset of considerable historic importance.

The cultural life of the city has expanded enormously in the past quarter-century. The new Art Gallery ranks among the top five in Canada. McMaster University has an art gallery as well, and there are a number of exhibition halls and art dealers. Theatre has experienced a somewhat slower expansion, perhaps because of the competition from touring shows, and from Toronto. Hamilton has had its own professional theatre for over a decade, and the older groups continue to put on successful plays and musicals. The university has a program in Dramatic Arts, and presents performances of drama and dance.

However, the greatest expansion has been in classical music. The region can boast of the Hamilton Philharmonic and McMaster Symphony, plus a number of excellent chamber orhcestras, quartets, trios, and soloists. The most remarkable demonstration of the love of music has been in the number and quality of choral groups which may have resulted partly because singing allows the easiest participation by the public at large, and partly because the heavy post-war immigration from the Netherlands and Great Britain has supplied the area with thousands of people accustomed to singing. In addition to the ten or so choral groups giving concerts, there are well-known industrial choirs, several excellent church choirs, and 'ethnic' singing groups. Some choirs are able to attract members from a wide radius; thus the Bach-Elgar Chorus includes singers from Toronto, Guelph, Brantford, and St Catharines. It is no exaggeration to maintain that Hamilton must be one of the principal choral centres of Canada, although none of the city's booster literature even refers to this wealth of singing.

The city has not made the same kind of mark in popular music. The emphasis on the big acts, the costs involved in producing the music, and the transiency of what's 'in' have placed participation beyond the range of possibility for most of the population. Yet, Hamilton does have its unique character and reputation; perhaps a 'Steel City' sound is waiting to be born.

This outburst of cultural development is one of the manifestations of the growing appreciation of diversity. Canada's self-definition as a 'mosaic' rather than a 'melting pot' has become commonly accepted only since the 1960s; it has granted a new degree of prominence to the non-British ethnic groups in the area. Hamilton has had a rich ethnic mix at least since 1900, but until the 1960s the non-British were psychologically as well as locationally shunted off to the northern and eastern margins of the built-up area. They have now become highly visible in their parades and demonstrations, costumes and folk dances, restaurants and feasts. The Italians and Germans in particular have burst out of their meeting halls and church basements to include the entire city in their celebrations of Festa Italia, Oktoberfest, and Karnaval. 'Ethnic' restaurants abound, notably those of the Chinese, Italians, and Greeks.

But even beyond the ethnic mix there is a fundamental diversity, almost a dichotomy, in the structure of the population. The city includes both the 'hard-nosed' industrial workers towards the east and the professionals and intellectuals towards the west. Each of these supports its own favoured set of activities. In some ways it is a schizophrenic city, with one end only dimly aware of what the other end is doing. If diversity increases interest, Hamilton must be one of the most fascinating cities in Canada.

Probably the most significant change of the past three decades has been the move away from the 'puritan' work ethnic towards a fresh appreciation of nature and of leisure. More than any other city, Hamilton has profited from this shift in ideology, because it was an ultimate example of an industrial city with an unmatched verge of nature around it. Pollution control and abatement has become a matter of universal concern. Air pollution has been lessened, particularly those emissions originating in the larger factories, but much remains to be done. Every autumn when a stagnant air mass, closed off by a shallow inversion lid, raises the pollution count over the dreaded figure of 32, we are reminded of how vulnerable an industrial area can be.

Water pollution remains of equal concern, particularly in the corners of the bay, the basins near the Red Hill and Spencer creeks. Here, too, much has been done; much remains to be done. Fortunately the prevailing perception of the function of the bay has changed. It is no longer seen only as a port and a source for industrial cooling; it is now beginning to be recognized as a multi-purpose resource with great possibilities for leisure activities. The western end is now clearly set aside for recreation and housing. It is already one of the finest boating harbours on Lake Ontario. The placement of the Canada Centre for Inland Waters, across the bay from the steel plants, has guaranteed continuing monitoring of the problems of the bay, continuing publicity, and continuing efforts towards a cleanup. Cootes Paradise, the marshy innermost end of the bay system, has become an integral part of the Royal Botanical Gardens, and has been flanked by a series of nature hiking trails. The Dundas end of 'the marsh' has even become the home of a large flock of Canada Geese.

The most satisfying result of the movement towards the enjoyment of the out-of-doors has been the reclamation of the unmatched peripheries of the urban area. The Bruce Trail has been laid out along the crest of the Niagara escarpment (the 'Mountain Brow') right through Hamilton, Stoney Creek, and Ancaster, and around Dundas. Parks and conservation areas occupy the crest for much of its length, and two downhill ski runs take advantage of the 100-metre (328-foot) drop. The Dundas Valley is filled with rolling hills piled up by the glacier between two wings of the escarpment; it is being organized, through selective purchase, into a magnificent conservation area, replete with trails for hiking in the summer and cross-country skiing in the winter. The Royal Botanical Gardens, with their extension around Cootes Paradise, have been made into one of the finest garden and nature-preserve complexes in Canada, and certainly the finest in the middle of an urban area. A waterfront park has been developed along Lake Ontario, although the temperature of the water guarantees that bathing in the lake is reserved for the hardy (or the foolhardy). Other parks and conservation areas allow for a more comfortable form of bathing, and for picnicking. Fed by the Fruit Belt to the east, and the apple orchards and berry patches to the northwest, the region resembles a cornucopia in the summer and autumn.

There is now a completeness to the region which can be highly satisfying to the permanent resident. The city is visibly rooted in the past, as witnessed by its preserved landmark buildings as well as the chronological mix of industrial structures. There is a fascinating mix of housing types, ethnic groups, and cultural and leisure pursuits. The completeness is fashioned too by the escarpment which encloses the lower city and most of the suburbs.

Hamilton is more of a lake port than any other Canadian city east of Thunder Bay, and every winter several of the 'lakers' are berthed at its docks. Over 90 per cent of its 9.1 million to 12.7 million tonnes (10 million to 14 million tons) of freight per year is directly related to the steel industries which occupy the eastern half of the waterfront. The remaining portion, small though it is relatively, nevertheless represents traffic with many of the countries of the world, via the St Lawrence Seaway. The present general cargo port is in virtually the same area where the original port developed over a century and a half ago.

There remain, of course, problems which must be faced as Hamilton moves towards the coming century. The eastern end of the downtown should be saved and brought back to prosperity. A continued expansion in employment possibilities is an obvious necessity which receives constant attention. Perhaps more subtle is the slow deterioration of the older housing stock north and east of the city centre. Hamilton cannot always depend on an influx of immigrants to supply the persons who will take over these houses and maintain them. In particular the belt north of King Street, between James and Gage, must be allowed to renew itself if the city is not to suffer the American malaise of the vacuum near the centre. Fundamental to this aim may be the removal of the trucks and speeding cars from Cannon and Wilson streets. Those idealistic people who are working to forestall any attempt to have heavy traffic move around the southern end of the city may, without realizing it, be guaranteeing the continuing increase in heavy traffic in the lower city, and the death of that area as a suitable residential zone. As suggested in other chapters, many of the social needs of the poor, the infirm, the ailing, will need to be met. There is also the continuing need to work on the cleanup of the environment, but this too is a matter which receives constant attention. Perhaps most difficult of all will be the changing of the image of Hamilton which is still common currency in the rest of Canada.

Images die hard, and the perception of Hamilton held by much of the country is still frozen into the city's reputation of thirty years ago. Perhaps Hamilton boasted too loudly of being first 'the Birmingham of Canada,' and later 'the Pittsburgh of Canada.' The labels stuck, and have continued to be reinforced by the skewed view seen from the Skyway Bridge. For those people who see only that view and never deign to enter the city, the image may remain ineradicable. The proud label of the 1950s has become the detraction of the 1980s, and its effects can be overcome only by enticing enough people, and especially the image makers, to come and see the city for what it now is.

REFERENCES

Freeman, B. 1976. Hamilton's civic square: the first eleven years. *City Magazine* 1 (no. 8): 26-41

McKay, A.G. 1967. *Victorian Architecture in Hamilton*. W.L. Griffin Ltd, Hamilton

Nader, G.A. 1976. *Cities of Canada*, vol. 2: *Profiles of Fifteen Metropolitan Centres*. Macmillan of Canada, Toronto

Weaver, J. 1982. *Hamilton: An Illustrated History*. James Lorimer and Company, Toronto

Index

Hamilton, City of: architecture 293-4; commercial space 254; Corktown area 108; culture 297; downtown redevelopment 290-3, 295; ethnic status 147-8; expansion 125, 134-5, 286, T17.2; family status 148-50, 152-3; future employment 250; future growth 246; initial survey 99; inner city 289-90; North End 108; population in 1851 108; in 1891 114; in 1951-81 286-7; in early twentieth century 122-3; port develops 103-4; preservation of buildings 296; recreation 299; social change (1961-81) 150-3; social conditions 107-8, 121, 125, 133; social structure 140-4; transport access 119; unemployment 262

Hamilton, economy: crash of 1857 111; decline in early 1980s 240, 262; early 107; early–twentieth-century 123; Great Depression 130; industrial structure 240; late–nineteenth-century 116; occupational structure 239; post Second World War expansion 134

Hamilton Aero club 130
Hamilton (Mount Hope) airport 253
Hamilton General Hospital 190
Hamilton Harbour 26, F2.7; infilling 115, 125, 216; Stelco and 202
Hamilton Harbour Commission, established 127
Hamilton Mountain: barrier to transport 120, 128; early settlement 127, 128; effects on climate 34-5, 175; future growth 248; slums on 133; social status 144; trunk sewer 242; see also Niagara Escarpment
Hamilton Psychiatric Hospital 190
Hamilton Region Conservation Authority 32

Hamilton-Wentworth, Regional Municipality of 156; attempts to secede from 166; creation 164; extent of 158-60; financial effects 165-6; growth 187; land use in 182; location 13; reasons for 157-8; Steele Commission 158-64; Stewart Commission 167; structure 160-1, 166

Hess Village 152
Hinkley, Alderman Brian 194, 196
housing conditions 128, 133; wartime 133
Hughson, Nathaniel 101, 104
Hydro lines, effects on swamps, 91
hydrology: disturbances 90; regime 29; water storage 88

immigrants, location of 139
immigration: 1820-40 106-7; 1970s 146, 153; nineteenth-century 106-7; effects of 107; first European 78
information society 261-3
Inglewood (house) 113
inversions. See climate
iron ore, sources of 214
Iroquois Bar 126, 129; see also Burlington Heights
Ivor Wynne Stadium 130

Jockey Club 136
Jolley Cut 15, 135

Keefer, Thomas 113

Lake Iroquois 15, 24
Lake Ontario: effect on climate 34; freezing 34; origin 25; regulation of levels 25; water levels 25
Lake Warren 23
Lake Whittlesey 23
Land, Robert 104